A Conservation Assessment
of the Terrestrial Ecoregions
of Latin America and the Caribbean

A Conservation Assessment of the Terrestrial Ecoregions of Latin America and the Caribbean

Eric Dinerstein
David M. Olson
Douglas J. Graham
Avis L. Webster
Steven A. Primm
Marnie P. Bookbinder
George Ledec

Published in association with The World Wildlife Fund

The World Bank
Washington, D.C.

ISBN 0-8213-3295-3

Library of Congress Cataloging-in-Publication Data

A conservation assessment of the terrestrial ecoregions of Latin
    America and the Caribbean / Eric Dinerstein . . . [et al.].
        p.   cm.
    Includes bibliographical references (p.   ).
    ISBN 0-8213-3295-3
    1. Biological diversity conservation—Latin America—Evaluation.
2. Biological diversity conservation—Caribbean Area—Evaluation.
3. Biotic communities—Latin America—Evaluation.   4. Biotic
communities—Caribbean Area—Evaluation.   5. Ecology—Latin America—
Evaluation.   6. Ecology—Caribbean Area—Evaluation.
I. Dinerstein, Eric, 1952–
QH77.L25C66   1995
333.9516'098—dc20   95-227
                                                                  CIP

Contents

Appendixes

Glossary   123

References   127

Figures

Tables

Maps

Foreword

As leading financiers of biodiversity conservation in Latin America and the Caribbean, the World Bank and the Global Environment Facility (GEF) have a responsibility to target conservation funds to the areas of greatest concern and need. The present study represents a step toward this goal by helping us to understand more clearly the conservation status of the terrestrial ecoregions of Latin America and the Caribbean.

Donors and conservation planners have traditionally focused their attention on the lowland rainforests, extremely species-rich habitats. The present study emphasizes that many other habitat types are also very important for biodiversity and merit close attention in conservation strategies. Tropical dry broadleaf forests, montane forests, coniferous forests, temperate forests, grasslands,

savannas, shrublands, and drylands are among the highest priorities for conservation action in Latin America and the Caribbean.

The information and orientation this report provides will support Bank operations that involve the protection or management of natural habitats. For the GEF, the results of this study will be useful at the regional planning level.

This study is also noteworthy because of the collaboration between the Bank and the World Wildlife Fund, a leading nongovernmental organization for conservation. The convergence of priorities between our two organizations suggests an evolving collaboration on conservation issues. I extend my thanks to the World Wildlife Fund for contributing its expertise and for working with Bank specialists to make this report possible.

Ismail Serageldin
Vice President
Environmentally Sustainable Development
The World Bank

Acknowledgments

This study has benefited from the contributions of many biodiversity specialists, representing an array of academic institutions, conservation organizations, and government agencies from Latin America and the Caribbean (LAC) or with strong interests in the region. These individuals and the organizations they represent are listed in the Contributors section.

Within LATEN, the World Bank's Environment Unit for Latin America and the Caribbean, we single out the role of Michel de Nevers, who originally proposed the idea for this study. Dean Gibson, Guillermo Paz y Miño, and Manuel Bonifaz all worked as LATEN research assistants on this project. Dennis Mahar and later William Partridge supervised this project. Peter Brandriss prepared the report for publication. Kathy Dahl and Tracey Smith from the Bank's Office of the Publisher provided very valuable editing and publishing assistance. Luis A. Solórzano translated the report into Spanish.

Substantial funding for this study was provided by the Global Environment Coordination Division of the World Bank, responsible for administering the Bank's Global Environment Facility (GEF) portfolio. We would like to highlight the support and contributions of Kenneth Newcombe and Kathy Mackinnon.

The World Wildlife Fund (WWF) contributed substantial funds of its own for the completion of this study. WWF, through a grant from the Ford Foundation, provided funding for the mangrove conservation assessment workshop. We extend our thanks to all of the mangrove experts who participated in the workshop. The LAC program of WWF provided valuable expertise in assessing the conservation status of ecoregions and their biological distinctiveness, and helped develop the decision rules for prioritizing among ecoregions. Josh Podowski, Stuart Sheppard, and Ian Heywood helped with the map digitizing and other related tasks. Pia Iolster, Emily Aikenhead, and Pedro Lopez-Valencia assisted with data research. Eric Dinerstein was supported by an Armand G. Erph Conservation Fellowship.

We would also like to thank the Biodiversity Support Program (BSP), supported by the United States Agency for International Development (USAID), for indirectly facilitating our work through their funding of a LAC conservation priority-setting exercise. The BSP exercise included a major workshop, attended by over 60 regional biodiversity specialists, which provided a significant amount of data for this study.

Some groups helped by supplying extensive spatial databases at no cost to this project. We are grateful for their generosity and their commitment to sharing conservation data. We acknowledge Chuck Carr, John Robinson, the Wildlife Conservation Society, and colleagues for the use of their data on remaining vegetation and protected areas of Central America. Frank Reichenbacher and David Brown of Southwestern Field Biologists provided a map of biotic communities of Central America and the Caribbean. Kent Redford and Roger Sayre of The Nature Conservancy provided data on several countries of LAC. The Colombian Centro Internacional para la Agricultura Tropical (CIAT) provided data on protected areas of LAC. Jorge Soberón and his colleagues at CONABIO (the Mexican Comisión Nacional para el Conocimiento y Uso de la Biodiversidad) provided data on the remaining vegetation of Mexico. Tom Stone and his colleagues at Woods Hole provided advanced very high resolution radiometer (AVHRR) data on remaining vegetation of South America. Gonzalo Castro and colleagues at Wetlands for the Americas provided maps of LAC wetlands. We also received valuable unpublished data from the national park agencies of Argentina and Bolivia. The Environmental Systems Research Institute, and in particular Jack Dangermond, Lance Shipman, and Charles Convis, provided technical support and made software available for

GIS applications. Hewlett-Packard, Inc., and Apple Computer, Inc., provided technical support through hardware and software donations. The World Conservation Monitoring Centre provided data on remaining vegetation and protected areas for selected countries.

A number of individuals reviewed early drafts of the methodology and shared their ideas. These included James Quinn, Gordon Orians, John Terborgh, Leonardo Lacerda, Kent Redford, Allan Putney, John Mackinnon, and Javier Simonetti. Several individuals read the final draft report and made valuable comments, including Annie Brunholzl, Kent Redford, Margaret Symington, Steve Cornelius, Garo Batmanian, Miguel Pellerano, Luis Solórzano, and Kathy Mackinnon.

Authors

Eric Dinerstein, Ph.D., Director, Conservation Science Program, World Wildlife Fund-US, Washington, D.C.

David M. Olson, Ph.D., Conservation Scientist, Conservation Science Program, World Wildlife Fund-US, Washington, D.C.

Douglas J. Graham, M.S., Environmental Specialist, Environment Unit, Technical Department, Latin America and the Caribbean Region, The World Bank, Washington, D.C.

Avis L. Webster, M.S., Spatial Technologies Specialist, Conservation Science Program, World Wildlife Fund-US, Washington, D.C.

Steven A. Primm, M.S., Research Associate, Conservation Science Program, World Wildlife Fund-US, Washington, D.C.

Marnie P. Bookbinder, M.S., Conservation Science Fellow, Conservation Science Program, World Wildlife Fund-US, Washington, D.C.

George Ledec, Ph.D., Ecologist, Environment Unit, Technical Department, Latin America and the Caribbean Region, The World Bank, Washington, D.C.

Contributors

E. Aikenhead (World Wildlife Fund-US)

J. Alcorn (Biodiversity Support Program, USA)

C. Alderman (World Bank, USA)

C. Alho (World Wildlife Fund–BRAZIL)

M. Ayres (Universidade Federal do Pará, BRAZIL)

P. Bacon (University of the West Indies, TRINIDAD AND TOBAGO)

J. Barborak (Wildlife Conservation Society, USA)

E. Barriga (USAID, COLOMBIA)

G. Batmanian (World Wildlife Fund–US)

M. Baudoin (Secretaría Nacional del Medio Ambiente, BOLIVIA)

M. Bonifaz (World Bank, USA)

P. Brandriss (World Bank, USA)

J. Brokaw (USAID, USA)

D. Brown (Southwestern Field Biologists, USA)

K. Brown (Universidade de Campinas [UNICAMP], BRAZIL)

D. Bryant (World Resources Institute, USA)

E. Bucher (Universidad Nacional de Córdoba, ARGENTINA)

G. Burgess (University of Florida, USA)

B. Buschbacher (Belém Project Office, World Wildlife Fund–US)

J. Cajal (Fundación para la Conservación de las Especies y Medio Ambiente [FUCEMA], ARGENTINA)

P. Canevari (Wetlands for the Americas, ARGENTINA)

M. Canevari (Administración de Parques Nacionales, ARGENTINA)

C. Carr (Wildlife Conservation Society, USA)

G. Castilleja (World Wildlife Fund–US)

G. Castro (Wetlands for the Americas, USA)

R. Cavalcanti (Universidade de Brasília, BRAZIL)

B. Chernoff (Field Museum of Natural History, USA)

K. Chomitz (World Bank, USA)

M. Cifuentes (World Wildlife Fund–US)

G. Cintrón-Molero (USFWS, USA)

M. Collins (World Conservation Monitoring Centre, UK)

J. E. Conde (Laboratorio de Biología Marina, Centro de Ecología y Ciencias Ambientales, VENEZUELA)

S. Contreras Balderas (Universidad Autónoma de Nuevo León, MEXICO)

C. Convis (Environmental Systems Research Institute, USA)

S. Cornelius (World Wildlife Fund–US)

L. Couto (Universidade Federal de Viçosa, BRAZIL)

J. Dangermond (Environmental Systems Research Institute, USA)

P. Dansereau (Université du Québec à Montréal, CANADA)

I. Davidson (Wetlands for the Americas, CANADA)

R. de la Maza (Conservation International, MEXICO)

P. DeVries (Harvard University, USA)

B. Dias (Ministério do Meio Ambiente e da Amazônia Legal, BRAZIL)

A. Dickie (USAID, USA/GUATEMALA)

L. Diego Gómez (Organization for Tropical Studies, COSTA RICA)

R. Dirzo (Universidad Nacional Autónomo de México [UNAM], MEXICO)

W. Duellman (University of Kansas, USA)

J. Echevarria (Universidad Nacional de Tumbes, PERU)

F. Erize (ARGENTINA)

C. Espinosa (Instituto Nicaragüense de Recursos Naturales y del Ambiente [IRENA], NICARAGUA)

E. Ezcurra (Secretaría de Desarrollo Social [SEDESOL], MEXICO)

R. Ford Smith (VENEZUELA)

F. Flores Verdugo (Centro Unidad Mazatlan en Agricultura y Manejo Ambiental, MEXICO)

G. Fonseca (Conservation International, BRAZIL)

E. *Forno G.* (Centro de Datos para la Conservación, BOLIVIA)

R. *Foster* (Field Museum of Natural History, USA)

P. *Foster-Turley* (USAID, USA)

P. *Freeman* (Environment and Natural Resources Information Center [ENRIC], USA)

C. *Freese* (World Wildlife Fund–US)

E. *Fuentes* (CHILE)

M. *García Donayre* (Instituto Nacional de Recursos Naturales [INRENA], PERU)

R. *Gauto* (Fundación Moises Bertoni, PARAGUAY)

D. *Gibson* (World Bank, USA)

W. M. *Graham* (CANADA)

A. *Grajal* (Wildlife Conservation Society, USA)

G. *Grau* (World Wide Fund for Nature–UK)

D. *Gross* (World Bank, USA)

D. *Heesen* (USAID, COSTA RICA)

A. *Henderson* (New York Botanical Gardens, USA)

O. *Herrera-MacBryde* (IUCN/Smithsonian, USA)

I. *Heywood* (World Wildlife Fund-US)

S. *Higman* (Guyana Forest Commission, GUYANA)

R. *Howard* (Arnold Arboretum, USA)

O. *Huber* (Fundación Botanico, VENEZUELA)

R. *Huber* (World Bank, USA)

E. *Iñigo-Elias* (University of Florida, USA)

P. *Iolster* (World Wildlife Fund-US)

A. *Iriarte Walton* (Servicio Agricola y Ganadero, Ministerio de Agricultura, CHILE)

J. *Izquierdo* (FAO, CHILE)

D. *Janzen* (University of Pennsylvania, USA/COSTA RICA)

N. *Johnson* (World Resources Institute [WRI], USA)

P. *Jones* (CIAT, COLOMBIA)

M. *Kalin de Arroyo* (Universidad de Chile, CHILE)

C. *Kane* (World Wildlife Fund-US)

S. *Keel* (The Nature Conservancy, USA)

S. *Kelleher* (Biodiversity Support Program, USA)

R. *Kerr* (The Hope Zoo, JAMAICA)

M. *Kiernan* (World Wildlife Fund–US)

B. *Kirmse* (World Bank, USA)

P. *Koohafkan* (FAO, ITALY)

M. *Kux* (USAID, USA)

L. *Lacerda* (World Wide Fund for Nature—International)

G. *Lamas* (Universidad Nacional Mayor San Marcos, PERU)

P. *Lopez-Valencia* (World Wildlife Fund-US)

R. *Luxmoore* (World Conservation Monitoring Centre, UK)

N. *Maceira* (Instituto Nacional de Tecnología Agropecuaria [INTA], ARGENTINA)

R. *Machado* (Conservation International, USA)

J. *Mackinnon* (Asian Bureau of Conservation, HONG KONG)

K. *Mackinnon* (World Bank, USA)

J. *Madill* (CANADA)

S. *Malone* (Conservation International, SURINAME)

F. *Marcoux* (CANADA)

J. *Mariaca P.* (Dirección Nacional de Conservación de la Biodiversidad, BOLIVIA)

P. *Marquet* (Pontificia Universidad Católica de Chile, CHILE)

J. *McKenna* (World Bank, USA)

G. *Medina* (World Wildlife Fund–US)

N. *Menezes* (Universidade de São Paulo, BRAZIL)

F. *Mereles* (Universidad de Asunción, PARAGUAY)

R. *Mittermeier* (Conservation International, USA)

B. *Moffat* (World Wildlife Fund–US)

A. *Molnar* (World Bank, USA)

A. *Moreira* (USAID, BRAZIL)

J. *Morello* (Harvard University, USA)

D. *Neill* (Fundación Jatun Sacha, ECUADOR)

D. *Nepstad* (Woods Hole Research Center, USA)

K. *Newcombe* (World Bank, USA)

S. *Oliver* (World Bank, USA)

S. *Olivieri* (Conservation International, USA)

G. *Orians* (University of Washington, USA)

J. *Ottenwalder* (United Nations Development Programme [UNDP], DOMINICAN REPUBLIC)

G. *Paz y Miño* (World Bank, USA)

M. *Pellerano* (World Wildlife Fund–US)

C. A. *Peres* (Universidade de São Paulo, BRAZIL)

T. *Pierce* (USAID, GUATEMALA)

J. *Pitt* (University of Grenada, GRENADA)

C. *Plaza* (World Bank, USA)

J. *Podowski* (World Wildlife Fund-US)

B. *Potter* (Island Resources Foundation, USA)

G. *Powell* (Centro Científico Tropical, COSTA RICA)

S. *Price* (World Wildlife Fund–CANADA)

G. *Prickett* (USAID, USA)

A. *Putney* (IUCN, USA)

L. *Quevedo* (World Wildlife Fund–BOLIVIA)

J. *Quinn* (University of California, USA)

K. *Redford* (The Nature Conservancy, USA)

H. *Reichart* (World Wide Fund for Nature–NETHERLANDS)

F. *Reichenbacher* (Southwestern Field Biologists, USA)

M. *Ribera* (Centro de Datos para la Conservación, BOLIVIA)

J. *Robinson* (Wildlife Conservation Society, USA)

M. *Rodrigues* (Universidade de São Paulo, BRAZIL)

C. *Rodstrom* (Conservation International, USA)

P. *Rosabal* (IUCN, PUERTO RICO)

J. *Ruiz* (ECOFONDO, COLOMBIA)

E. *Santana* (Universidad de Guadalajara, MEXICO)

C. *Saravia T.* (Centro de Investigaciones Ecológicas del Chaco, ARGENTINA)

F. *Sarmiento* (University of Georgia, USA)

K. *Saterson* (Biodiversity Support Program, USA)

R. *Sayre* (The Nature Conservancy, USA)

Y. Schaeffer-Novelli (Universidade de São Paulo, BRAZIL)

S. Schonberger (World Bank, USA)

F. Seymour (World Wildlife Fund–US)

S. Sheppard (World Wildlife Fund–US)

L. Shipman (Environmental Systems Research Institute, USA)

J. Simonetti (Universidad de Chile, CHILE)

J. Soberón (CONABIO, MEXICO)

C. Sobrevila (World Bank, USA)

O. Solbrig (Harvard University, USA/ARGENTINA)

L. Solórzano (Princeton University, USA/COLOMBIA)

T. Stone (Woods Hole Research Center, USA)

D. Stotz (Field Museum of Natural History, USA)

M. Symington (Biodiversity Support Program, USA)

J. Terborgh (Duke University, USA)

K. Thelen (FAO, CHILE)

J. Tosi (Centro de Ciencias Tropicales, COSTA RICA)

R. Tuazon (Inter-American Development Bank, USA)

R. Twilley (University of Southwestern Louisiana, USA)

M. Vales G. (CUBA)

B. Watson (The Nature Conservancy, NICARAGUA/USA)

M. Webb Records (World Bank, USA)

R. Welch (SOS Mata Atlântica, BRAZIL)

C. Wicks (World Wide Fund for Nature–UK)

B. Wilcox (Institute for Sustainable Development, USA)

E. Wilcox (World Wildlife Fund–US)

N. Windevoxhel (IUCN Regional Wetlands Programme for Mesoamerica, COSTA RICA)

D. Wood (World Wildlife Fund–US)

M. Yates (USAID, BOLIVIA)

E. Yerena O. (Instituto Nacional de Parques [INPARQUES], VENEZUELA)

F. Zadroga (USAID, MEXICO)

L. Zeitlin-Hale (University of Rhode Island, USA)

Acronyms and Abbreviations

AVHRR	Advanced very high resolution radiometer
BSP	Biodiversity Support Program
CI	Conservation International
CIAT	Centro Internacional para la Agricultura Tropical (Colombia)
CONABIO	Comisión Nacional para el Conocimiento y Uso de la Biodiversidad (Mexico)
FAO	Food and Agriculture Organization of the United Nations
GEF	Global Environment Facility
GIS	Geographic information system
IUCN	The World Conservation Union
LAC	Latin America and the Caribbean
LATEN	Environment Unit (Technical Department) for LAC (The World Bank)
MET	Major ecosystem type (see also Glossary)
MHT	Major habitat type (see also Glossary)
MSS	Multispectral scanner
TNC	The Nature Conservancy
USAID	United States Agency for International Development
USFWS	United States Fish and Wildlife Service
WRI	World Resources Institute
WWF	World Wildlife Fund

Executive Summary

This priority-setting study elevates, as a first principle, maintaining the representation of all ecosystem and habitat types in regional investment portfolios. Second, it recognizes landscape-level features as an essential guide for effective conservation planning. Without an objective framework to assess the conservation status and biological distinctiveness of geographic areas, donors run the risk of overlooking areas that are seriously threatened and of greatest biodiversity value.

The lack of such an objective regional framework prompted this study, whose goals were (a) to replace the relatively ad hoc decisionmaking process of donors investing in biodiversity conservation with a more transparent and scientific approach; (b) to move beyond evaluations based largely on species lists to a new framework that also incorporates maintaining ecosystem and habitat diversity; (c) to better integrate the principles of conservation biology and landscape ecology into decisionmaking; and (d) to ensure that proportionately more funding be channeled to areas that are of high biological value and under serious threat.

The study was financed by the World Bank, the Global Environment Facility, and the World Wildlife Fund (WWF), and it was carried out by WWF's Conservation Science Program in close collaboration with the World Bank's Environment Unit for Latin America and the Caribbean.

The study's biogeographic approach to setting conservation priorities begins by dividing LAC into 5 major ecosystem types (METs), 11 major habitat types (MHTs), and 191 ecoregions. This last number includes 13 mangrove complexes but otherwise excludes marine areas and freshwater habitats (except large, geographically contiguous freshwater ecosystems).

The classification scheme for ecoregions builds on existing work and, wherever possible, uses ecoregion boundaries recognized by conservation planners and biogeographers. The classification scheme helps maintain the goal of representing all habitat and ecosystem types in an investment portfolio. The scheme also allows us to assess conservation status using criteria tailored to the distinct dynamics and spatial patterns of biodiversity of different ecosystem types.

The conservation assessment integrates two fundamental data layers: *conservation status* and *biological distinctiveness*. To assess the conservation status of ecoregions, we adapted the Red Data Book approach used by the World Conservation Union (IUCN) for species. We conducted a snapshot conservation assessment of ecoregions to classify them as either Extinct (completely converted), Critical, Endangered, Vulnerable, Relatively Stable, or Relatively Intact. The classifications are based on five indicators of landscape integrity: total loss of original habitat, number and size of blocks of intact habitat, rate of habitat conversion, degree of fragmentation or degradation, and degree of protection. We assumed that unfavorable changes in these indicators lowered the probability of ecological processes and major components of biodiversity being maintained within a given ecoregion. The snapshot conservation status is further modified to a final conservation status after considering potential threats over the next 20 years to ecoregions based on their type, timeframe, spatial scale, and intensity.

The biological distinctiveness of an ecoregion is assessed within its MHT, thus ensuring that we compare tropical moist broadleaf forests only to other tropical moist broadleaf forests and not to montane grasslands with major differences in patterns of biodiversity. We classify ecoregions as Globally Outstanding, Regionally Outstanding, Bioregionally Outstanding, or Locally Important.

To undertake these analyses, we solicited help from a wide range of biodiversity specialists and conservation planners from the LAC region. Regional experts reviewed the landscape level criteria and threats in order to determine the conservation status. They also provided data and assessments of the key variables used to determine the biological distinctiveness of ecoregions: beta diversity, species richness and endemism for several major taxa, unique ecological communities or processes, and rarity or distinctiveness of ecosystem or habitat types at varying biogeographic scales.

As indicated by the final conservation status of the 178 non-mangrove ecoregions, we identified 31 ecoregions as Critical, 51 as Endangered, 55 as Vulnerable, 27 as Relatively Stable, 8 as Relatively Intact, and 6 as Unclassified. Thirty-nine ecoregions were considered more threatened after the assessment of threat was applied to the snapshot conservation status. The highest number of Critical and Endangered ecoregions occurred in tropical moist broadleaf and tropical dry broadleaf forests. However, only 3 percent of the tropical dry broadleaf forest ecoregions were Relatively Stable or Relatively Intact, whereas 27 percent of the tropical moist broadleaf forest ecoregions were Relatively Stable or Relatively Intact, indicating that tropical dry broadleaf forests are on average more threatened than any other forest type in the LAC region. Most (83 percent) xeric ecoregions were either Critical, Endangered, or Vulnerable.

Thirty-four ecoregions in LAC were deemed Globally Outstanding and 32 others were considered Regionally Outstanding. There were 59 Bioregionally Outstanding ecoregions and 47 Locally Important ecoregions. Numerically, tropical moist broadleaf forests contain most of the Globally Outstanding ecoregions and 47 percent of all the Regionally Outstanding ecoregions. They are concentrated in the western portion of Amazonia and the tropical northern Andes, with others scattered among northern Mexico, the Atlantic forests of Brazil, and southern Chile. Montane grasslands have the highest proportion of Globally or Regionally Outstanding ecoregions, followed by tropical moist broadleaf forests. Three MHTs (tropical dry broadleaf forests; grasslands, savannas, and shrublands; and deserts and xeric shrublands) had a higher proportion of Bioregionally Outstanding and Locally Important ecoregions than did the other MHTs. All MHTs were represented by at least one ecoregion classified as either Globally or Regionally Outstanding.

The integration of the conservation status of ecoregions with their biological distinctiveness helps to identify biodiversity conservation priorities. We used a simple matrix of these two variables to identify the ecoregions of highest conservation importance. Each MHT was analyzed separately in order to maintain representation of all major habitat types. Fifty-five out of 178 ecoregions distributed among the 10 non-mangrove MHTs were designated as of Highest Priority at the Regional Scale (level I). These include 23 ecoregions in tropical moist broadleaf forests; 5 in tropical dry broadleaf forests; 2 in temperate forests; 5 in tropical and subtropical coniferous forests; 2 in grasslands, savannas, and shrublands; 4 in flooded grasslands; 8 in montane grasslands; 2 in Mediterranean scrub; 2 in deserts and xeric shrublands; and 2 in restingas.

We also created and applied a set of decision rules designed to achieve better bioregional representation. Applying these decision rules added 19 ecoregions to the highest priority list.

Mangrove units were assessed separately; we assigned priority rankings to the 40 units making up the 13 mangrove complexes. Rankings were assigned to investments for restoration, conservation with restricted use, or conservation for sustainable use.

Patterns of biodiversity and landscape integrity of whole ecoregions are given precedence over other analytical layers because changes in these two layers are essentially irreversible. The integration model based on these two layers identifies priorities and suggests the timing and sequence of investment and the level of effort required to conserve biodiversity.

The matrix and decision rules presented can help guide governments and donors in preventing complete degradation and conversion in the most threatened regions. Such investments may be quite expensive. The matrix also identifies the most intact examples of biologically outstanding ecoregions with the best chance for long-term persistence and the highest potential for cost-effective investments. We stop short of identifying specific investment priorities within ecoregions, which should be done only after assessing political, social, and economic factors. These secondary factors are more fluid than biological variables and, in our view, are best applied in intra-ecoregion analyses. Application at finer geographic scales will better direct scarce funds to the sites within ecoregions that are the most intact, biologically important, and most likely to preserve biodiversity over the long term.

Introduction

The tropical, subtropical, and temperate habitats of Latin America and the Caribbean (LAC) contain some of the most important areas in the world for biodiversity conservation. Many of these areas face severe threats, and financiers and donors have responded to these pressures by financing biodiversity conservation programs across the region. The World Bank and the Global Environment Facility (GEF) have invested heavily in projects that support the creation or strengthening of protected areas, institutional strengthening of parks departments, protection of endangered species, and management of natural resources.[1] The total portfolio of conservation investments of the World Bank and the GEF, particularly as new GEF projects for LAC are approved over the next few years, probably exceeds the financial contributions of all other conservation donors combined.

If targeted properly, the resources of the GEF, the World Bank, and other major financiers and donors can have enormous impact in stemming the destruction and degradation of biologically important areas in LAC. Specialists within the GEF and LATEN, the World Bank's Environment Unit for Latin America and the Caribbean, recognized three years ago the need for better information to help target these resources. This study was one way of addressing these concerns. The World Bank contracted the Conservation Science Program of the World Wildlife Fund (WWF) to carry out a study to identify geographic priorities for biodiversity conservation based on the integration of the biological distinctiveness and conservation status of ecoregions.

A second objective was to create an approach that would be rigorous enough to be incorporated into the conservation planning exercises of LAC countries and other donors active in the region. The approach developed for this study has indeed influenced several priority-setting exercises underway in LAC:

- A priority-setting exercise financed by USAID, coordinated by the Biodiversity Support Program (BSP), and conducted by the WWF Conservation Science Program, the Wildlife Conservation Society, The Nature Conservancy, World Resources Institute, Conservation International, and a large group of regional experts. This exercise was designed to assist USAID in identifying geographic areas of high priority for biodiversity investments in LAC (BSP 1995).
- A priority-setting exercise conducted by the World Wildlife Fund's LAC program as part of its strategic planning activities to design the next three-year funding cycle (Wood and Cornelius 1994).
- Argentina's National Parks Department's analysis of the conservation status of Argentine ecoregions using methods based in part on a preliminary draft of our method (Olson and Dinerstein 1994). The analysis of Cajal (1994) is similarly based on an early draft of the method.

The ecoregion base map developed for this study is being widely used, even prior to publication, in a number of studies and planning activities. These include: an analysis of indigenous areas by Wilcox and Duin (1995); The Nature Conservancy planning exercise for their Parks in Peril program; and a priority-setting exercise for LAC wetlands conducted by Wetlands for the Americas.

---

1. As of July 1995, the World Bank had an active portfolio of 21 approved loans, to most of the major countries of LAC, that specifically included resources targeted for biodiversity conservation. These loans represent a total investment in biodiversity of about $470 million of which about $320 million is direct Bank lending. The World Bank is also one of the implementing agencies of the GEF, and in LAC, five World Bank GEF biodiversity projects had been approved by July 1995 for a total of $72 million. A sum of about $290 million has been pledged to the G-7 Pilot Program to Conserve the Brazilian Rain Forest, coordinated by the World Bank, which has many projects now in the planning or implementation stages.

This exercise, and other similar efforts, seek to advance biodiversity conservation planning beyond previous approaches, such as "Hotspots" (Myers 1988) and the "Megadiversity-country" approach (Mittermeier and Werner 1990). In these approaches, geographic priorities were selected largely on the basis of the single indicator of species richness. "Hotspots" and "Megadiversity" approaches did not directly address what is now widely accepted as a central goal of regional and national biodiversity conservation strategies: to maintain representation of all ecosystem and habitat types in order to conserve the distinct communities of organisms they contain (Noss 1992; Scott et al. 1993; Noss and Cooperrider 1994; Dinerstein et al. 1994). This study, and others like it, signals a shift in priority-setting focus, away from species richness as the sole or major discriminator, and towards conserving ecosystem and habitat diversity.

Although building on excellent previous work, we highlight eight specific advances of this study:

- A hierarchical classification scheme that divides LAC into major ecosystem types (METs), major habitat types (MHTs), and 191 ecoregions to ensure that representation is maintained at an appropriate biogeographic scale for regional conservation planning
- A transparent method that incorporates the principles of landscape ecology to assess the snapshot conservation status of the ecoregions of LAC
- A classification of the conservation status of ecoregions in the tradition of the IUCN Red Data Books
- An assessment of conservation status and threats tailored to the particular dynamic features and threats specific to different types of ecosystems
- An approach for assessing the biological distinctiveness of ecoregions at varying biogeographic scales
- Consideration of mangrove ecosystems, ecologically and economically valuable formations that have received inadequate attention from conservation donors
- Well-documented GIS databases that can be used by other individuals to reanalyze and update the data layers in this analysis
- A method for integrating biological distinctiveness of ecoregions with their conservation status designed to identify regional biodiversity priorities for conservation in each MET and to promote bioregional representation.

We limit our analyses to terrestrial ecoregions. A regional priority-setting initiative for freshwater eco-

systems is currently being undertaken by Wetlands for the Americas. The World Bank, in collaboration with the Great Barrier Reef Marine Park Authority and the The World Conservation Union (IUCN) have recently published a four volume series on global priorities for marine ecosystems (Kelleher et al. 1995). These exercises complement this effort.

This report is divided into seven chapters. The first chapter (Approach) provides an overview of the methodologies used in this study. A more detailed description of the methods can be found in Appendixes A and B. The second chapter provides a description of the METs and MHTs found in LAC, their geographic locations, and the ecoregions they contain. See more detailed definition of MHTs in Appendix C and the hierarchical classification scheme of ecoregions in Appendix D. Brief descriptions of the ecoregions and key sources consulted for their delineation, classification and assessment are in Appendix F.

The third chapter presents the results of the analysis of conservation status by MHT and across MHTs. The raw data for each ecoregion used to compute conservation status are in Appendix E and detailed information on sources consulted for remaining natural habitat and protected areas is in Appendix G. The fourth chapter presents an assessment of the biological distinctiveness of ecoregions within their MHTs. The fifth chapter explains how to use the methodology as a tool in conservation planning. The sixth chapter addresses the conservation status of mangroves of LAC which we assessed using an approach adapted from that developed for the terrestrial ecoregions.

The last chapter provides our conclusions and recommendations. After summarizing the results of the study, we discuss the possibility of using our approach at an ecoregional or intra-ecoregional scale. Such analyses at finer geographic scales are an essential follow-up to this study. Without them, donors run the risk of financing biodiversity conservation in the most important ecoregions within a major habitat type, but conserving some of the less important habitat blocks within those ecoregions, or at an insufficient level to conserve their biodiversity. We offer some guidelines that highlight maintaining landscape integrity and better conserving areas characterized by high rates of species turnover (beta diversity), and we give prominence to unique habitats.

It is essential that biodiversity be conserved everywhere, but clearly some areas have higher value than others and degrees of threat vary. The geographic biodiversity priorities identified by this study take these considerations into account. They also serve as a precursor for guiding *investment priorities*, which can only be determined after economic, social, institutional and political factors are considered within priority ecoregions. This study stops short of determining site-specific investment priorities, which, in any event, should be determined by comprehensive planning efforts conducted at the ecoregion or country level. Nonetheless, we stress that biodiversity assessments and analyses of landscape integrity should be considered the foundations of priority-setting exercises at any scale.

Comments on this report would be welcomed by the World Bank or the World Wildlife Fund at the following addresses:

- Douglas J. Graham or George Ledec
  Environment Unit, Technical Department
  Latin America and the Caribbean Region
  The World Bank
  1818 H Street, NW
  Washington, DC 20433 USA
  Fax: (202) 676-9373
  Internet: DGraham@Worldbank.org, or
     GLedec@Worldbank.org

- Eric Dinerstein
  Conservation Science Program
  World Wildlife Fund
  1250 24th Street, NW
  Washington, DC 20037-1175 USA
  Fax: (202) 861-8377

1

Approach

Fundamental Goals Underlying the Approach

The terrestrial ecoregions of Latin America and the Caribbean (LAC) cover the spectrum—from some of the driest deserts, such as the Atacama of Chile and Peru, to some of the wettest forests, such as the Chocó/Darién forests of Panama, Colombia, and Ecuador. A few of these ecoregions are almost completely converted or degraded, whereas others contain vast, intact blocks of original habitat. The degree of protection of remaining habitat blocks also varies widely from one ecosystem type to another. In some cases there is nearly complete protection, while in others there is little or no protection either now or in the foreseeable future. Most of the ecoregions in LAC fall between these extremes.

This study provides a biologically based framework for a conservation assessment of LAC's diverse ecoregions. The design framework, based on widely accepted goals in conservation biology and landscape ecology (Noss 1992; Krever et al. 1994), incorporates a number of analytical steps to help determine biodiversity conservation priorities (Figure 1-1). The fundamental goals underlying our approach include:

- Representation of all distinct natural communities
- Maintenance of ecological and evolutionary processes that create and sustain biodiversity
- Maintenance of viable populations of species
- Conservation of blocks of natural habitat large enough to be responsive to large-scale periodic disturbances and long-term changes (e.g., global warming).

Goal 1: Representation

Maintaining representation is critical to conserving ecosystem and habitat diversity. Also, conservation biologists are in virtual consensus that conservation of representative examples of the most diverse array of natural habitats in a region is the most cost-effective way to prevent species extinction. Previous efforts at priority-setting that were largely determined by species lists fail to capture the diversity of ecosystems and the biota they contain, to the detriment of xeric and non-forested formations in particular (Scott et al. 1993; Csuti in prep).

To achieve the goal of representation, we developed a hierarchical classification scheme that divides LAC into 5 major ecosystem types (METs), 11 major habitat types (MHTs), and 191 ecoregions (Figure 1-2). The hierarchy helps achieve representation by assessing priorities at biologically meaningful levels. Chapter 2 presents in more detail LAC's METs, MHTs, and ecoregions.

An ecoregion is defined as a geographically distinct assemblage of natural communities that share a large majority of their species, ecological dynamics, and similar environmental conditions and whose ecological interactions are critical for their long-term persistence. We believe that the ecoregion unit is the minimum level of resolution required for achieving regional representation and effective conservation planning.[1] In this manner, ecoregions that fall under tropical moist broadleaf forests, for example, are compared to other such forests rather than to coniferous or grassland ecoregions.

Ecoregions within the same major habitat type can be similar in their structure and ecological processes but share few species due to biogeographic barriers or species turnover with distance. Representation of biodiversity at the level of species—as well as at higher hierarchical levels of ecosystems and habitat types—is critical in any comprehensive regional

---

1. The boundaries of an ecoregion should identify an area over which a single conservation strategy could be effectively applied.

4

Figure 1-1. Analytical Steps Used to Derive Biodiversity Conservation Priorities

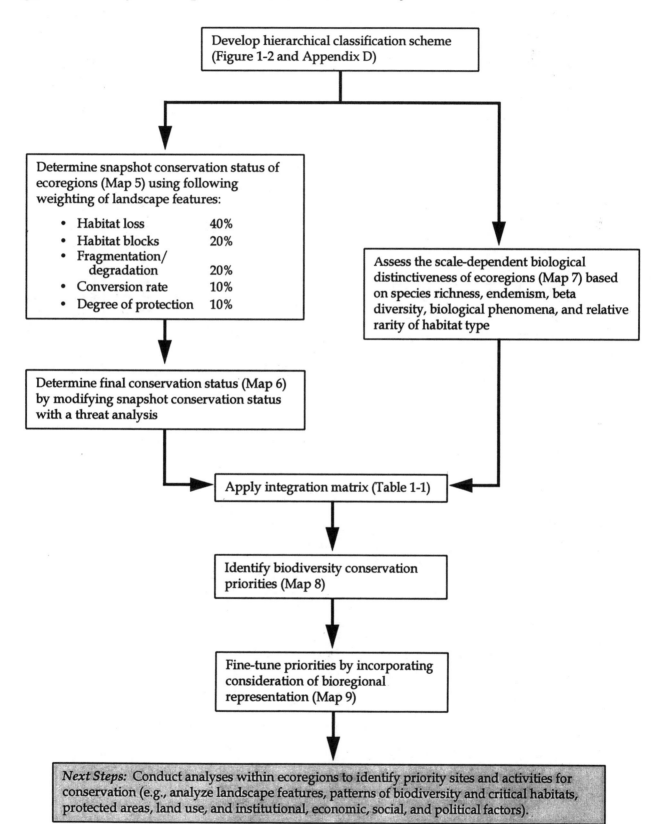

Figure 1-2. Hierarchical Classification Scheme of METs, MHTs, Ecoregions, and Bioregions

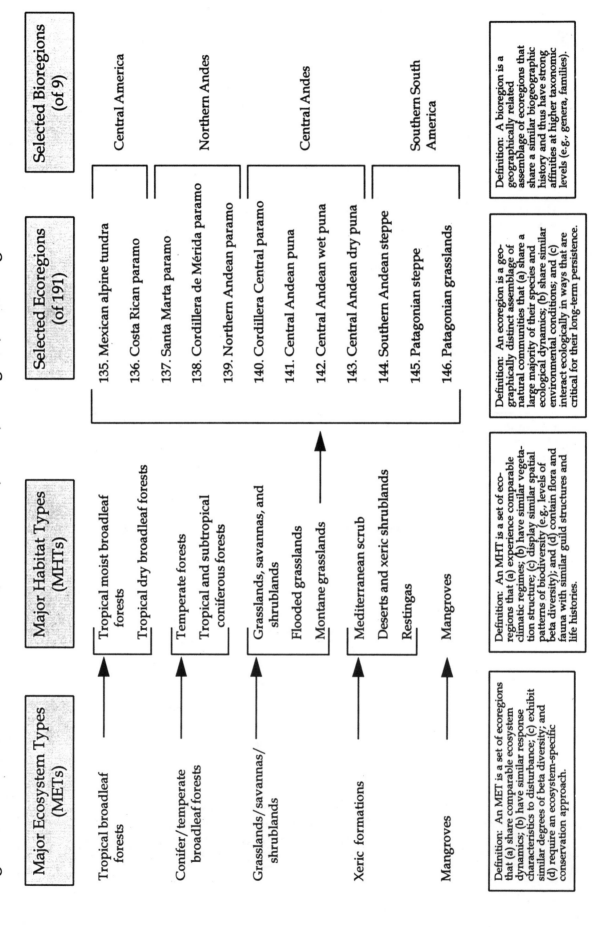

biodiversity strategy. For these reasons, representation must be achieved among ecoregions that belong to the same MHT but occur in different geographic areas of LAC.

We thus defined nine discrete biogeographic areas that we call bioregions. As explained in more detail in subsequent chapters, these bioregions are used both in the definition of scale-dependent biological distinctiveness and to fine-tune the final results of the priority-setting exercise to achieve the goal of better biogeographic representation. The nine bioregions, which reflect the biogeographic distinctiveness of major areas of LAC (Map 1), are Caribbean, Northern Mexico, Central America, Orinoco, Amazonia, Northern Andes, Central Andes, Eastern South America, and Southern South America.

Goals 2 to 4: Maintenance of Ecological Processes and Viable Species Populations

Directly assessing the status of ecological processes and viability of species populations (Goals 2 and 3) is often logistically difficult, but estimates are critical for a sound priority-setting framework. We achieve this goal with an index of conservation status using landscape-level features. The indicators are the loss of original habitat, the presence of large blocks of contiguous habitat, degree of fragmentation and degradation, rates of habitat conversion, and the degree of protection within an ecoregion. A key assumption of our approach is that the indicators selected, and the manner in which they are weighted, serve as robust predictors of (a) the extent to which important components of biodiversity (e.g., rare keystone species or the presence of viable populations of top predators) will likely persist over time; and (b) the maintenance of important ecological processes. The method to determine conservation status is discussed further in this chapter and in Appendix A.

Maintaining areas large enough to withstand periodic, large-scale disturbances and long-term changes (Goal 4) is also best assessed using landscape-level features such as the presence of large blocks of original habitat and total loss of original habitat. Degradation, indicated by these landscape features, would have major impacts on regional biodiversity conservation because extensive loss of original habitat and large blocks of natural habitats is essentially irreversible. With few exceptions—such as some mangrove units and some savannas—restoration of converted or highly degraded habitats is too slow and costly to be an effective conservation tool. Thus, achieving Goal 4 depends on sufficiently large, relatively intact units that still function naturally.

Snapshot Conservation Status

The conservation status of ecoregions was determined by summing the numerical values assigned to the five key landscape-level variables mentioned above: loss of original habitat, number and size of large blocks of original habitat, degree of fragmentation and degradation, rate of conversion of remaining habitat, and degree of protection. To assess landscape-level features, we relied on expert opinion and on spatial databases and maps (where such data were available and of reasonable quality). More detailed information on assessment of landscape-level features by ecoregion is presented in Appendix A, and information on remaining vegetation and protected areas is in Appendix G.

To what extent are the ecoregions that support populations, communities, and unique ecosystems threatened? To answer this question, we assessed conservation status of ecoregions in the tradition of the IUCN Red Data Books. The IUCN Red Data Book categories of threat have gained widespread acceptance as a framework for determining the conservation status of species and populations. Such assessments have been codified into various Red Data Books to call attention to species and populations considered to be on the verge of extinction (IUCN 1988; Collar et al. 1992). This study marks the first time that ecoregions have been classified this way.

Mace and Lande (1991) reviewed IUCN categories of threat and suggested three objectively defined categories: Critical, Endangered and Vulnerable. Inspired by this review, we classified ecoregions as Extinct (completely converted), Critical, Endangered, Vulnerable, Relatively Stable, and Relatively Intact. Classification was determined by weighting the numerical values defined for each of the five landscape variables listed above. In weighting these variables, we gave much greater prominence to loss of original habitat and number of large blocks of intact habitat. We believe that these variables are the best indicators of the probability of persistence of ecological processes within ecoregions. The ranges of values used to classify the ecoregions were derived from the conservation biology, theoretical ecology, and landscape ecology literature (see Appendix A).

In each of the definitions below, we first describe the ecoregion's landscape integrity as assessed by the more quantifiable criteria listed in Appendix A. The remaining sentences describe qualitatively the predicted ecological and biological impacts of loss of landscape integrity. They reflect how with increasing habitat loss, degradation, and fragmentation, ecological processes cease to function naturally, or at all, and major components of biodiversity are steadily eroded. The categories of conservation status are:

- *Extinct.* No natural communities resembling the original ecosystems remain. Some of the original biota are still present but persist only within highly modified communities and landscapes. No opportunities for restoration of the original natural communities exist due to permanent alteration of physical conditions, loss of source pools of native species, substantive alteration of natural ecological processes, or an inability to eradicate or control aggressive alien species.
- *Critical.* The remaining intact habitat is restricted to isolated small fragments with low probabilities of persistence over the next 5-10 years without immediate or continuing protection and restoration. Many species are already extirpated or extinct due to the loss of viable habitat. Remaining habitat fragments do not meet the minimum area requirements for maintaining viable populations of many species and ecological processes. Land use in areas between remaining fragments is often incompatible with maintaining most native species and communities. Spread of alien species may be a serious ecological problem, particularly on islands.
- *Endangered.* The remaining intact habitat is restricted to isolated fragments of varying size (a few larger blocks may be present) with medium to low probabilities of persistence over the next 10-15 years without immediate or continuing protection or restoration. Some species are already extirpated due to the loss of viable habitat. Remaining habitat fragments do not meet minimum area requirements for most species populations and large-scale ecological processes. Land use in areas between remaining fragments is largely incompatible with maintaining most native species and communities. Top predators are almost exterminated.
- *Vulnerable.* The remaining intact habitat occurs in habitat blocks ranging from large to small; many intact clusters will likely persist over the next 15-20 years, especially if given adequate protection and moderate restoration. In many areas, some sensitive or exploited species have been extirpated or are declining, particularly top predators and game species. Land use in areas between remaining fragments is sometimes compatible with maintaining most native species and communities.
- *Relatively stable.* Natural communities have been altered in certain areas, causing local declines in exploited populations and disruption of ecosystem processes. These disturbed areas can be exten-

sive, but are still patchily distributed relative to the area of intact habitats. Ecological linkages among intact habitat blocks are still largely functional. Guilds of species that are sensitive to human activities, such as top predators, the larger primates, and ground-dwelling birds, are present but at densities below the natural range of variation.
- *Relatively intact.* Natural communities within an ecoregion are largely intact with species, populations, and ecosystem processes occurring within their natural ranges of variation. Guilds of species that are sensitive to human activities, such as top predators, the larger primates, and ground-dwelling birds, occur at densities within the natural range of variation. Biota move and disperse naturally within the ecoregion. Ecological processes fluctuate naturally throughout largely contiguous natural habitats.

The criteria for classifying ecoregions were tailored to reflect biological and ecological differences among METs. To produce the snapshot assessment, we relied upon regional experts at the BSP workshop (BSP/CI/TNC/WRI/WWF 1995), the mangrove workshop, the WWF LAC program biogeographic priorities workshop (Wood and Cornelius 1994), the World Bank, and elsewhere to assess the landscape-level criteria. Where data were up to date, we used overlays of remaining vegetation to assess landscape-level features for forested MHTs. We look forward to other specialists updating and refining this analysis as higher quality data become available.

Final Conservation Status

The "snapshot assessment" of conservation status incorporated an estimation of threat in the sense that rate of conversion and degree of degradation and fragmentation were considered. However, one can easily identify other severe threats that are likely to affect the longer-term trajectory of conservation efforts in a particular ecoregion. The type, scale, intensity, and timeframe of threats (e.g., deforestation, mining, overgrazing, pollution, overharvesting of wildlife) need to be assessed to make conservation status a more reliable planning tool (refer to Appendix A for more details). Snapshot conservation status and threat assessments—to give a final conservation status for each ecoregion—were analyzed by the same organizations and individuals.

Biological Distinctiveness

Assessing the relative biological importance of ecoregions must be part of any comprehensive priority-setting exercise. For this study, we interpret the biological importance of ecoregions as the degree to which its biodiversity (both components and processes) is distinctive at different biogeographic scales. We define this scale-dependent assessment as the biological distinctiveness of an ecoregion. All ecoregions are biologically distinct to some degree, particularly at the level of species and species assemblages and at increasingly broader biogeographic scales. However, some ecoregions are so exceptionally rich, complex, or unusual that they merit extra attention from conservation planners.

In order to ensure appropriate biodiversity comparisons, the biological distinctiveness of an ecoregion is only assessed within its MHT (e.g., montane grasslands were compared only to other ecoregions of this MHT). Four categories of biological distinctiveness at different biogeographic scales are used in this study: Globally Outstanding, Regionally Outstanding, Bioregionally Outstanding, and Locally Important. Ecoregions are classified as outstanding if—for the biogeographic scale being considered—they have extraordinary levels of the attributes described in either criteria 1, 2, 3, or 4, or are evaluated as meeting either criteria 5 or 6 described below:

1. Species richness with an emphasis on the following taxa: plants, mammals, birds, reptiles, amphibians, and butterflies.[2] For example, the Napo region of western Amazonia is considered a Globally Outstanding ecoregion because of having the world's highest known alpha diversity for many taxa.
2. Endemism (i.e., the number and proportion of species occurring only in that ecoregion) with an emphasis on the same taxa as species richness.
3. Complexity of species distributions within the ecoregion (e.g., beta diversity, gamma diversity at larger scales, and patterns of local endemism).
4. Uniqueness and rarity of certain ecological phenomena in terms of their structure or dynamic properties. For example, on a global scale, Varzea forests, with their remarkable seasonal migrations of fish to flooded forests, and the Galapagos Islands, a xeric scrub ecoregion renowned for extraordinary adaptive radiations, are classified as Globally Outstanding.
5. Number of ecoregions in the same MHT. An ecoregion is considered Globally Outstanding if less than seven ecoregions in its MHT exist

in the world (e.g., Mediterranean scrubs), Regionally Outstanding if less than three occur regionally, and Bioregionally Outstanding if it is the only example of its MHT in its bioregion (e.g., Paraguaná restingas, which are the only example of coastal dune formations in northern South America).

6. Size of the ecoregion. The largest example(s) of an MHT at global and regional scales are characterized as outstanding because they maintain biodiversity processes and components characteristic of that habitat type, but are not always represented in smaller units. For example, the Pantanal is rated Globally Outstanding because it is one of the world's largest seasonally flooded freshwater complexes.

The biological distinctiveness assessments used in this study represent a first attempt to categorize ecoregions in this way. Rather than give quantitative weightings to the criteria described above, we sought a consensus of opinion from experts.[3] We recognize that measuring and assigning relative values to such a complex ecoregion attribute requires a number of subjective assessments, a task made even more challenging by the incompleteness of biodiversity data for many regions and taxa, by the fact that available data are not now systematically and comprehensively organized, and by the lack of global ecoregion maps of comparable scale and classification (like the one developed for LAC in this study).

However, we believe that the conservation community has access to sufficient information on continental patterns of biodiversity (through expert opinion and the technical literature) to identify ecoregions that are exceptionally distinctive at global, regional, and bioregional scales. As new data become available, some ecoregions might shift up or down one level. In general, we have most confidence in our classification of Globally Outstanding or Regionally Outstanding ecoregions.

Several MHTs were particularly challenging to assess because detailed biodiversity information or regional experts were unavailable during this study. Specifically, some tropical dry broadleaf forests, flooded grasslands, montane grasslands, and deserts

---

2. These taxa are the ones primarily assessed by the experts at the BSP workshop, but many other taxa were considered through further consultation with experts and review of the technical literature.

3. At the BSP workshop (BSP/CI/TNC/WRI/WWF 1995), regional habitat units—essentially aggregations of our ecoregions—were ranked as being of high, medium, or low biological importance, based primarily on species richness and endemism but also, depending on the taxa, taking into account phyletic diversity and rare and endangered species. To determine our own biological distinctiveness rankings, we reanalyzed the BSP workshop data with input from experts at the WWF-US LAC workshop (Wood and Cornelius 1994) and at the World Bank.

and xeric shrublands were difficult to evaluate. We expect and look forward to biogeographers revising ecoregion rankings, particularly as new biodiversity information becomes available.

All ecoregions not classified as outstanding are of local importance even though they are of average or less than average biological distinctiveness when compared with other ecoregions of the same MHT and bioregion. Nevertheless, it should be stressed that even Locally Important ecoregions harbor globally unique biodiversity, maintain important ecosystem services, and are useful to local human communities (Woodwell 1995). Such ecoregions should figure prominently in national or subregional biodiversity strategies.

Biodiversity Conservation Priority

Conservation status and biological distinctiveness are two essential discriminators for conservation priority-setting at regional scales (ecological function, a third important discriminator, is most effectively applied in finer-scale analyses). As a practical tool for conservation planning, biological distinctiveness assesses the relative rarity of different natural communities and phenomena that, in conjunction with other parameters, can be used to estimate the extent of opportunity for their conservation. Conservation status represents an estimate of the current ability of an ecoregion to maintain viable species populations, sustain ecological processes, and be responsive to short- and long-term change—three basic goals of biodiversity conservation. Moreover, patterns of biodiversity are fixed and large-scale landscape changes are largely irreversible within the timeframe of current conservation efforts. With the exception of some mangrove systems and a few other habitats, restoration of ecosystems is either impossible, too costly, or too slow to offer much scope for conservation investment. Similarly, exterminated species can never be recreated. The populations they comprise, or the assemblages, communities, and ecological processes they are a part of, are also difficult and costly to replace. In contrast, institutional, economic, social, and political discriminators show much more fluidity (e.g. reversibility) and are more subject to change within relevant time scales.

The Integration Matrix

We offer a simple integration matrix to help identify priority ecoregions for biodiversity conservation (Table 1-1). Along the horizontal axis, we arrange ecoregions by their final conservation status. Along the vertical axis, we classify ecoregions by their bio-

logical distinctiveness. To achieve representation among all habitat types, a separate matrix is created for each of the MHTs. The matrix allows us to classify each ecoregion into four biodiversity conservation priority categories (levels I-IV).

The following three considerations determine the selection of the seven "level I" cells:

- The relative rarity worldwide or the extraordinary biodiversity found in ecoregions designated as Globally Outstanding warrant their inclusion as "level I." Exceptions are ecoregions classified as Relatively Intact, which are ranked as "level II." These ecoregions are assumed to be under much less threat over the next several decades and do not warrant proportionately greater attention at this juncture.
- The certainty of severe and immediate loss of biodiversity in Critical and Endangered ecoregions classified as Globally Outstanding or Regionally Outstanding. Critical and Endangered ecoregions that are Bioregionally Outstanding or Locally Important are certainly important and contain unique species and communities; however, the integration model proposed here emphasizes that proportionately more attention should be given to ecoregions with the same conservation status but with higher biological distinctiveness.
- The opportunity to conserve biologically outstanding ecoregions that are on the precipice of major decline calls for inclusion of the top two Vulnerable cells. They are included to help guard against Vulnerable ecoregions slipping into Critical or Endangered status.

We recognize that alternative integration models may prioritize ecoregion attributes that are different than those used in our model (e.g., all Critical ecoregions) or may introduce new variables or modifiers such as the importance of different ecoregions in maintaining large-scale ecological functions such as global carbon sequestration or watershed capacity. The relative importance of different parameters may vary depending upon the spatial and temporal scale of conservation investments (e.g., urgent measures versus strategic conservation planning) and the kinds of activities under consideration.

Ensuring Bioregional Representation

The matrix presented on the following page provides, by MHT, a set of Highest Priority at Regional Scale ecoregions (level I) for all of LAC. However, this list within each MHT does not necessarily ensure bioregional representation.

Table 1-1. Matrix for Integrating Biological Distinctiveness and Conservation Status to Assign Priorities for Biodiversity Conservation

Biological Distinctiveness	*Final Conservation Status*				
	Critical	*Endangered*	*Vulnerable*	*Relatively Stable*	*Relatively Intact*
Globally Outstanding	I	I	I	I	II
Regionally Outstanding	I	I	I	II	III
Bioregionally Outstanding	II	II	III	III	IV
Locally Important	III	III	IV	IV	IV

Note: The roman numerals indicate biodiversity conservation priority classes:

Level I = Highest Priority at Regional Scale (shaded area)
Level II = High Priority at Regional Scale
Level III = Moderate Priority at Regional Scale
Level IV = Important at National Scale

Bioregional representation is important because of the dissimilarities in the species and natural communities of ecoregions within the same MHT but from different biogeographic zones of LAC; geographic replacement of species occurs as range boundaries, environmental gradients, and physical barriers are crossed at large geographic scales. By including bioregional representation as a goal in a regional biodiversity strategy, problems of scale are also dealt with in a more objective fashion. For example, many Caribbean ecoregions that share the same MHT with continental ecoregions come out poorly in priority rankings. Typically, small island faunas and floras are depauperate compared to continental biotas because of size and isolation. Ideally however, at least one example of each MHT in the Caribbean should be considered as Highest

Priority at Regional Scale to reflect the importance, at a LAC scale, of the biodiversity of that bioregion.

We propose that within each bioregion each constituent MHT be represented—subject to certain conditions—by at least one ecoregion classed as Highest Priority at Regional Scale. We attain this goal by "upgrading," when necessary, a single level II or level III ecoregion (but never a level IV) to "level Iᵃ." We choose a level II over a level III ecoregion and a worse conservation status over a better conservation status. Only level III ecoregions that are Critical, Endangered, or Vulnerable are considered eligible. If there is a tie between two or more ecoregions, we choose the ecoregion with the highest biological distinctiveness (either as ranked in Appendix E or, if ranked similarly, according to our judgment).

2

Major Ecosystem Types, Major Habitat Types, and Ecoregions of LAC

One of the major stumbling blocks to creating a rigorous framework for setting priorities at the regional level has been the absence of a widely accepted classification scheme of biogeographic units. In its absence, donors have often set priorities using political rather than biogeographic boundaries, even though the two bear little overlap. When donors use country boundaries they seldom consider the goal of maintaining representation of all ecosystem types and instead largely rely on methods based on species richness by country or region. The emphasis on species richness as an indicator of priority ecoregions has skewed interest to tropical moist broadleaf forests and caused us to neglect the diverse ecosystems and biota found in the drier, non-forested or semi-forested ecoregions of LAC (Redford et al. 1990).

Ignoring the goal of representation has also prompted some biologically misleading questions, such as, Which are more important: the paramo of Colombia, the Atlantic forests of Brazil, or the dry forests of Costa Rica? The question is inappropriate because it compares ecoregions representing markedly different habitat types. More appropriate questions would be, Among the montane grasslands of Central and South America, which are the most threatened and biologically important?, or Among the tropical moist broadleaf forests of South America, which are the most intact or threatened?

To answer such questions requires a hierarchy based on habitat types. This chapter and the maps at the back of this book present such a hierarchy. On the maps, we associate colors with habitat types. For example, xeric formations are coded in shades of red, grasslands/savannas/shrublands in shades of yellows and orange, tropical dry broadleaf forests in earth tones, conifer/temperate broadleaf forests in shades of blue, and tropical moist broadleaf forests in shades of green. Mangrove units are pink. Ecoregions on the maps represent estimates of the "original" pre-Colombian distribution of habitats and thus cover the entire LAC region.

Major Ecosystem Types (METs)

The first level of the hierarchy is the major ecosystem type (MET). We identify five METs: tropical broadleaf forests; conifer/temperate broadleaf forests; grasslands/savannas/shrublands; xeric formations; and mangroves (Figure 1-2). Ecoregions in a given MET tend to share (a) minimum area requirements for conservation of ecological processes and other components of biodiversity; (b) response characteristics to major disturbance; and (c) similar levels of beta diversity (the rate of turnover of species along elevational or environmental gradients). Thus, METs are defined primarily on dynamic properties and spatial patterns of biodiversity and not wholly on vegetation structure.

Tropical broadleaf forests cover the largest areas of LAC (9.29 million km²), followed by grasslands/savannas/shrublands (7.13 million km²), xeric formations (1.85 million km²), conifer/temperate broadleaf forests (1.09 million km²), and mangroves (40,623 km²). We do not dwell on METs in this report because they are too coarse for regional conservation planning. The next hierarchical level is that of the major habitat type (MHT). For each MET and MHT, defining characteristics, sensitivity to disturbance, and general conservation guidelines are described in Appendix C.

Major Habitat Types (MHTs)

MHTs represent habitat types that are similar in terms of their general structure, climatic regimes,

Table 2-1. Important Attributes of the Major Habitat Types of Latin America and the Caribbean

Major Ecosystem Type	Major Habitat Type	Total Size (km²)	Percent of LAC	Number of Ecoregions	Mean Ecoregion Size (km²)
Tropical broadleaf forests	Tropical moist broadleaf forests	8,214,285	38.0	55	149,351
	Tropical dry broadleaf forests	1,043,449	4.8	31	33,660
Conifer/temperate broadleaf forests	Temperate forests	332,305	1.5	3	110,768
	Tropical and subtropical coniferous forests	770,894	3.6	16	48,181
Grasslands/savannas/ shrublands	Grasslands, savannas, and shrublands	7,058,529	32.7	16	441,158
	Flooded grasslands	285,530	1.3	13	21,964
	Montane grasslands	1,416,682	6.6	12	118,057
Xeric formations	Mediterranean scrub	168,746	0.8	2	84,373
	Deserts and xeric shrublands	2,276,136	10.5	27	84,301
	Restingas	34,975	0.2	3	11,658
Mangroves	Mangroves	40,623	0.2	[See Chapter 6]	

major ecological processes, and level of species turn-over with distance (beta diversity), with flora and fauna showing similar guild structures and life histories. Nested within the 5 METs are 11 MHTs (Table 2-1); tropical moist broadleaf forests are most extensive and several xeric formation MHTs (e.g., restingas) are the smallest (Maps 2a-c).

General distribution of these MHTs is as follows:

- Tropical moist broadleaf forests dominate in much of southern Mexico and Central America, northern and northwestern South America, along the middle and lower slopes of the northern and central Andes, and along a narrow belt of eastern Brazil (the Atlantic moist forests). The largest areas of tropical moist broadleaf forests in the Caribbean occur in the Greater Antilles, particularly on Hispaniola and Cuba. Some experts further subdivide this MHT into lowland and montane forests with a demarcation at about 1,500 m. This is by far the largest MHT in original extent, and it contains the highest number of ecoregions as well as some of the largest.
- Tropical dry broadleaf forests predominate in southern Mexico, parts of Pacific Central America, the Greater Antilles, central and eastern Brazil, eastern Bolivia, and along parts of northwestern

South America. Tropical dry broadleaf forests are the second most numerous set of ecoregions, some of which are among the smallest.

- Temperate forests are restricted to the Southern cone of South America and include one of the world's seven temperate rain forests. Temperate forest ecoregions are rare in LAC but quite large in mean size.
- Tropical and subtropical coniferous forests are widely dispersed. The greatest concentrations occur in central and southern Mexico, northern Central America, southern Brazil, and the Greater Antilles.
- Grasslands, savannas, and shrublands are most conspicuous in the southern cone region of South America, Mexico, and the Llanos of Venezuela and Colombia. This is the second largest MHT in LAC, largely because we classified the Cerrado and the Chaco as savannas rather than as dry forests.
- Flooded grasslands are most prominent in Amazonia and central South America.
- Montane grasslands are found primarily along the Andes. From a global perspective, the LAC region contains the vast majority of the world's paramo-type vegetation—a rare ecoregion type. This category also includes cold temperate grasslands such as the Patagonian steppe.

- The Mediterranean scrub MHT shows a disjunct distribution, with one unit in northwestern Mexico/southwestern USA and the other in Chile. Although small in overall area, these units are of global importance because only five Mediterranean-type ecoregions occur in the world, and two are in LAC.
- Deserts and xeric shrublands are most extensive in Northern Mexico, parts of the Greater and Lesser Antilles, Colombia, Venezuela, and Pacific coastal South America (Peru, Chile).
- Restingas and coastal dune formations are largely restricted to Brazil and Argentina, though to a lesser degree they are also found in Venezuela and Mexico.
- Mangroves are limited to the tropical zone along the coasts of continental landmasses or on islands of the Caribbean and Pacific (Map 4). Mangroves were classified into 13 major mangrove complexes and further subdivided into 40 mangrove units. Mangroves were classified into these units based on hydrographic, geomorphic, and ecological features.

Ecoregions

Ecoregions represent geographically discrete units of major habitat types. The ecoregion boundaries shown in the ecoregion maps provided (Map 3 and large-format map) represent the estimated original extent of each unit of MHT. The ecoregion is the principal unit we use to analyze priorities. The derivation of these units is discussed in greater detail below.

An ecoregion represents a geographically distinct assemblage of natural communities that share a large majority of their species, ecological dynamics, and similar environmental conditions, and whose ecological interactions are critical for their long-term persistence. On the basis of this definition, some ecoregions that contain a mosaic of distinct habitat types (e.g., the Cerrado of Brazil with its grasslands, gallery forests, dry forests, and open woodland) are considered as a single ecological unit. We assume that conversion or degradation of any major portion of an ecoregion would alter the ecological processes and species population dynamics of remaining areas, but have relatively less impact outside the ecoregion's boundaries.

Delineating the terrestrial ecoregions of LAC proved to be an ambitious two-year undertaking. The original extent of LAC ecoregions was delineated on a base map using the Digital Chart of the World, and then digitized and organized in a geographic information system (GIS) database using ARC/INFO software. For more details on delinea-

Figure 2-1. Number of Ecoregions by Size Categories

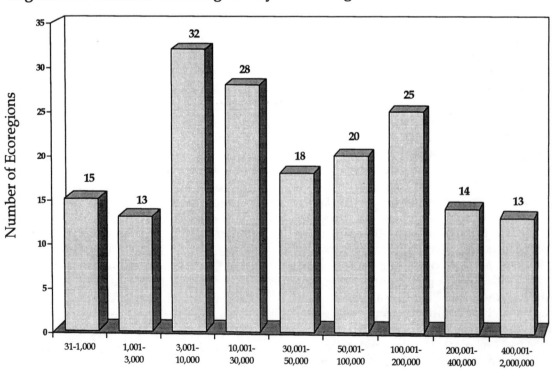

Ecoregion Size Category (km²)

tion of ecoregions, descriptions, and references, see Appendix F.

In most cases, the ecoregions are based on classifications from existing ecological studies in order to (a) enable this conservation assessment to benefit from the wealth of scholarly work that has been carried out in the region, and (b) to enhance the efficacy of conservation investments by working with classification systems that are widely used within the region. However, differences among regional systems are common, and we highlight here some of the decision rules we established for delineating boundaries.

Some ecoregions are composed of several disjunct areas because of bioclimatic factors, soils, hydrographic conditions, or isolation by physical barriers such as large rivers or mountain ranges. Such formations are considered as a single unit for the purposes of this analysis if they are geographically clustered (e.g., Tepuis) or maintain some level of biotic interaction among the habitat blocks. Some ecoregions and habitat types contain a continuum of ecological communities and classifying some of these systems was sometimes problematic. Examples include several partially wooded systems such as grasslands that contain smaller amounts of gallery forests, savannas, woodlands, and scrublands. In classifying these

ecoregions we emphasized the habitat that plays the most important role in maintaining ecological processes.

Ecoregions range in size from the Cerrado of Brazil at nearly 2 million km^2 to the Costa Rican paramo of only 31 km^2 (Appendix E). A breakdown by size class (Figure 2-1) shows that ecoregions vary enormously in size. Among the non-mangrove ecoregions, only 6.7 percent are less than 1,000 km^2, most of which are island units in the Caribbean bioregion. Ecoregions in the two smallest size classes are about equal in number to those in the two largest size classes. Among the four forested MHTs (tropical moist broadleaf forests, tropical dry broadleaf forests, temperate forests, and tropical and subtropical coniferous forests), tropical dry broadleaf forests are clustered around the smallest size categories and temperate forests—except for the Chilean winterrain forests—among the larger units. Grasslands/savannas/shrublands units are widely scattered. Flooded grasslands include the Pantanal—at nearly 141,000 km^2 one of the world's largest wetlands—and some much smaller units elsewhere in LAC. Other wetland units have been identified by Wetlands for the Americas and will be described in a forthcoming publication.

3

Conservation Status of Terrestrial Ecoregions of LAC

Conservation of biodiversity is important everywhere. However, in those ecoregions where human activities have caused widespread habitat destruction, the costs and level of effort to conserve biodiversity will be far higher than in more intact ecoregions. Some ecoregions are going to require the immediate implementation of recovery plans if most of the plant and animal communities they contain are to persist into the next decade. Other ecoregions will remain relatively stable over the next few decades.

Determining which ecoregions belong on a critical list requires weighing a number of factors, including their natural resiliency to major disturbance, their degree of beta diversity, the configuration of the remaining habitat, the degree of protection, and the type, intensity, and timeframe of major threats to biodiversity. For example, in most situations mangroves regenerate far more quickly from habitat disturbance than do tropical moist forests. Highly fragmented landscapes will normally be more threatened than those that contain large blocks of original habitat. Intensive grazing may be a less severe threat in grasslands than in dry forests. Ecoregions characterized by high beta diversity may require more protected areas that are well-distributed across the landscape to conserve the full complement of endemic plants and animals.

We provide an analytical framework for assessing the conservation status of ecoregions that is tailored to the particular dynamics of their respective major ecosystem type. A more detailed account of the methods we employed to evaluate conservation status is found in Appendix A, which also includes a more thorough discussion of the landscape-level features assessed in this methodology and a rationale for the weighting scheme.

The original extent of ecoregions map (Map 3 and large format map) illustrates what LAC probably looked like in pre-Colombian time (recognizing that some pre-Colombian landscapes were already anthropogenically modified to some degree). The snapshot conservation status assessment (Map 5) (i.e., an assessment based on current habitat configurations) provides insights into how seriously some ecoregions have been degraded. The presence of large blocks of original habitat, percent of remaining habitat, and degree of protection highlight opportunities for conservation within the ecoregion. Combined with rate of conversion and degree of degradation and fragmentation, these variables also help predict, from a biological perspective, the maintenance of ecological processes (e.g., predation, pollination and seed dispersal systems, nutrient cycling, migration, dispersal, and gene flow) that ultimately determine how much biodiversity will persist over the long term.

The final conservation status—the snapshot assessment of conservation status modified by threat (Map 6)—forecasts an ecoregion's trajectory given current trends over the next 5 to 20 years. The threat assessment is essential because the snapshot assessment, although valuable, only shows what presently remains. The threat assessment takes into account factors that we know or assume will impact biodiversity conservation in an ecoregion. For example, using our methodology, the snapshot assessment of the Guianan moist forests (tropical moist broadleaf forest MHT) indicates it is Relatively Intact. However, foreign timber concessions have already staked claim to large tracts of this ecoregion. With widespread logging and road building into pristine areas about to begin—and likely resulting in extensive habitat destruction and degradation—the snapshot assessment is changed to Relatively Stable. The final conservation status is used for integration with biological distinctiveness to determine conservation priorities (Chapter 5).

We begin by presenting the results of the snapshot conservation status of ecoregions. We then show how a threat analysis modifies the status, and we make comparisons within and across MHTs. Within an MHT, we identify the most critical ecoregions. Comparisons across MHTs allow us to determine if one MHT is more threatened than another. Finally, we highlight major trends in relationships between the size of ecoregions, the threats they face, and their biogeographic location.

Results

Snapshot Conservation Status

The most striking observation from the snapshot conservation status map of ecoregions (Map 5) is how few ecoregions are designated as either Relatively Stable (n=37; 22 percent) or as Relatively Intact (n=14; 8 percent). Essentially, much of the area in these two categories is located in the Amazon basin, the moist forests of the Petén (Mexico, Guatemala), and northwestern Mexican xeric systems. Twenty-three ecoregions (14 percent) are Critical, 47 (28 percent) are Endangered, and 45 (27 percent) are Vulnerable. Twelve ecoregions were not classified for lack of reliable information.

Critical and Endangered ecoregions are overrepresented in the tropical dry broadleaf forests, and in deserts and xeric shrublands (Table 3-1). By contrast, tropical moist broadleaf forests spanned a wide range but contained significantly more Relatively Stable and Relatively Intact ecoregions than did tropical dry broadleaf forests. The Relatively Stable and Relatively Intact tropical moist broadleaf forests include many of the largest ecoregions. From a broad bioregional perspective, Amazonia contains some of the most intact ecoregions. The other eight bioregions show about the same distribution of status classes.

Three ecoregions (Tehuantepec savannas, Veracruz palm savannas, and Veracruz pine-oak forests), either partly or entirely in Mexico, were classified as Extinct by some workshop experts but as Critical by others. We chose to designate all of these ecoregions as Critical for the final assessment.

Final Conservation Status

Final conservation status was derived by applying a threat analysis to the snapshot conservation status (half of the 12 previously unclassified ecoregions were also classified at this stage). Thirty-nine ecoregions were considered more threatened after the assessment of threat was applied to the snapshot conservation status. When analyzed by major habitat type, these break down to 23 (42 percent of the MHT) tropical moist broadleaf forest ecoregions;

Table 3-1. Snapshot Conservation Status of Ecoregions by Major Habitat Type

| Major Habitat Type | Snapshot Conservation Status | | | | | |
	Critical	Endangered	Vulnerable	Relatively Stable	Relatively Intact	Unclassified
Tropical moist broadleaf forests	3	13	16	13	10	
Tropical dry broadleaf forests	10	14	4	2		1
Temperate forests		1	1	1		
Tropical and subtropical coniferous forests	1	5	5	4	1	
Grasslands, savannas, and shrublands	2	1	6	5		2
Flooded grasslands	2	3	1	3	1	3
Montane grasslands			5	6		1
Mediterranean scrub		2				
Deserts and xeric shrublands	3	7	7	3	2	5
Restingas	2	1				
TOTAL	23	47	45	37	14	12

five (16 percent) tropical dry broadleaf forest ecoregions; one (33 percent) temperate forest ecoregion; two (13 percent) tropical and subtropical coniferous forest ecoregions; two (13 percent) grassland, savanna, and shrubland ecoregions; one (8 percent) flooded grassland ecoregion; three (25 percent) montane grassland ecoregions; one (50 percent) Mediterranean scrub ecoregion; and one (4 percent) desert and xeric shrubland ecoregion. Only one ecoregion (Puerto Rican moist forests) was considered less threatened after the assessment of threat was applied to the snapshot conservation status because of expected regeneration of natural forest.

The final conservation status of ecoregions revealed that 48 percent of all ecoregions are either Critical (18 percent) or Endangered (30 percent). Thirty-two percent are Vulnerable, 16 percent are Relatively Stable, and 5 percent are Relatively Intact (Table 3-2; Map 6). There are six ecoregions for which a final assessment could not be made because data were lacking. These are primarily xeric ecoregions of northern Mexico, which are currently being assessed under the Mexico Country Study for the Biodiversity Convention.

Numerically, the highest number of Critical and Endangered ecoregions occurred in tropical moist

Table 3-2. Final Conservation Status of Ecoregions by Major Habitat Type

| Major Habitat Type | Final Conservation Status | | | | | |
	Critical	Endangered	Vulnerable	Relatively Stable	Relatively Intact	Unclassified
Tropical moist broadleaf forests	6	15	19	11	4	
Tropical dry broadleaf forests	11	17	2	1		
Temperate forests		1	2			
Tropical and subtropical coniferous forests	3	3	5	4	1	
Grasslands, savannas, and shrublands	2	2	6	4		2
Flooded grasslands	3	4	3	2	1	
Montane grasslands			9	3		
Mediterranean scrub	1	1				
Deserts and xeric shrublands	3	7	9	2	2	4
Restingas	2	1				
TOTAL	31	51	55	27	8	6

Table 3-3. Final Conservation Status by Size of Ecoregion

| Size of Ecoregion (n=number of ecoregions) | Final Conservation Status | | | | | |
	Critical	Endangered	Vulnerable	Relatively Stable	Relatively Intact	Unclassified
31-1,000 km^2 (n=15)	3	5	3	3	1	
1,001-3,000 km^2 (n=13)	3	3	4	1	2	
3,001-10,000 km^2 (n=32)	8	9	10	4		1
10,001-30,000 km^2 (n=28)	6	11	5	3		3
30,001-50,000 km^2 (n=18)	6	7		3	1	1
50,001-100,000 km^2 (n=20)	2	7	9	1	1	
100,001-200,000 km^2 (n=25)	1	5	13	5		1
200,001-400,000 km^2 (n=14)	2	2	4	4	2	
400,001-2,000,000 km^2 (n=13)		2	7	3	1	

broadleaf forests and tropical dry broadleaf forests (Table 3-2). However, only 3 percent of the tropical dry broadleaf forest ecoregions were Relatively Stable or Relatively Intact, whereas 27 percent of the tropical moist broadleaf forests were Relatively Stable or Relatively Intact. From these data, we can conclude that tropical dry broadleaf forests are on average more threatened than either tropical moist broadleaf forests or conifer/temperate broadleaf forests in the LAC region. The restingas MHT contains the most threatened ecoregions among MHTs, followed by the Mediterranean scrub MHT. Most xeric formation ecoregions were either Critical, Endangered, or Vulnerable. Grassland, savanna, and shrubland ecoregions illustrated the widest spread in conservation status. In general, montane grasslands are less threatened than are lowland grasslands. There was no clear relationship between size of original extent of an ecoregion and its final conservation status (Table 3-3).

As a means to test the results of the threat-modified assessment, we asked the participants in the WWF LAC strategic planning workshop to repeat the threat-modified assessment without having access to the BSP assessments. Overall the two workshops yielded similar rankings, with the WWF LAC team reaching different conclusions from the BSP workshop for only 26 of the 178 non-mangrove ecoregions (14 percent).

Among the nine bioregions, the Northern Andes had the highest percentage (69 percent) of ecoregions clustered in the two most threatened categories (Table 3-4). Central America (67 percent), the Caribbean (57 percent), Eastern South America (50 percent), and Northern Mexico (39 percent) also had significant proportions of their ecoregions in the most threatened categories.

Table 3-4. Final Conservation Status by Bioregion

Bioregion	Critical	Endangered	Vulnerable	Relatively Stable	Relatively Intact	Unclassified
Northern Mexico	3	4	4	4	3	6
Central America	11	11	5	5	1	
Caribbean	2	11	9	1		
Orinoco		4	3	3	2	
Amazonia	1	3	7	11	2	
Northern Andes	9	11	8	1		
Central Andes	1	3	7			
Eastern South America	4	2	6			
Southern South America		2	6	2		

4

Biological Distinctiveness of Terrestrial Ecoregions of LAC at Different Biogeographic Scales

Ecoregions vary by the number of species they contain, by the level of endemism, and by the uniqueness of the assemblages, natural communities, ecological interactions, and biological phenomena found within them. In defining categories of biological distinctiveness, we included the relative rarity of certain MHTs besides the more commonly used variables of species richness and endemism. For example, some MHTs are represented by only a few ecoregions worldwide. Mediterranean scrub, temperate rain forests, and paramo are three such types that are rare globally but prominent in LAC.

We also pay greater attention to a less-used but important criterion, the degree of beta diversity of an ecoregion. Simply put, beta diversity is a measure of the turnover of species with distance or along environmental gradients. Several areas of LAC, such as the tropical Andes, contain areas of extraordinary beta diversity, where major turnover of species assemblages is common over a small elevational change or across nearby mountain ranges. By contrast, one can travel for hundreds of kilometers within other ecoregions and find similar assemblages of plants and animals.

The conservation of areas high in beta diversity, and how to predict their occurrence at finer geographic scales is now a major new thrust of the science of conservation biology and, we predict, will soon become an essential component in future priority-setting exercises.

We confine comparisons of biological distinctiveness to ecoregions within the same MHT, link biological distinctiveness to a biogeographic spatial scale, and state explicitly the rules for assigning ecoregions to one of the four distinctiveness classes (see Chapter 1). The specific reasons for designating ecoregions as either Globally or Regionally Outstanding are presented in Appendix F. The biodiversity of every ecoregion has unique elements and is ecologically important and this approach highlights the value of protecting each ecoregion in a national biodiversity strategy. From a regional perspective, some ecoregions have a higher level of biological distinctiveness than others, and they merit greater attention from conservation planners.

Results

Thirty-four ecoregions in LAC were deemed Globally Outstanding and 31 ecoregions were considered Regionally Outstanding (Map 7; Table 4-1). There are 62 Bioregionally Outstanding ecoregions and 50 Locally Important ecoregions.

Globally and Regionally Outstanding Ecoregions

An analysis of the biological distinctiveness ratings of ecoregions by MHT shows that, numerically, Globally Outstanding ecoregions are most commonly found in tropical moist broadleaf forests (Table 4-1). These ecoregions are concentrated in the western portion of Amazonia and the tropical Andes (Map 7).

If this MHT were further subdivided into lowland and montane forests, the Globally Outstanding units would be divided almost equally. The other examples are scattered among northwestern Mexico, the Atlantic forest in Brazil, and central and southern Chile.

Table 4-1. Biological Distinctiveness of Ecoregions by Major Habitat Type

	Biological Distinctiveness			
Major Habitat Type	*Globally Outstanding*	*Regionally Outstanding*	*Bioregionally Outstanding*	*Locally Important*
Tropical moist broadleaf forests	16	13	18	8
Tropical dry broadleaf forests	3	2	7	19
Temperate forests	1	1	1	
Tropical and subtropical coniferous forests	2	6	4	4
Grasslands, savannas, and shrublands	1	1	9	5
Flooded grasslands	1	3	8	1
Montane grasslands	4	4	3	1
Mediterranean scrub	2			
Deserts and xeric shrublands	2	1	12	12
Restingas	2		1	
TOTAL	34	31	63	50

In addition, tropical moist broadleaf forests contain 42 percent of the Regionally Outstanding ecoregions. Among the other MHTs, montane grasslands have the highest proportion (67 percent of the MHT) of Globally or Regionally Outstanding ecoregions, followed by tropical moist broadleaf forests (56 percent of the MHT). Tropical dry broadleaf forests; grasslands, savannas, and shrublands; and deserts and xeric shrublands had the lowest proportion of Globally or Regionally Outstanding ecoregions, but all MHTs were represented by at least one ecoregion classified as either Globally or Regionally Outstanding.

5

Integrating Biological Distinctiveness and Conservation Status

The integration of the conservation status of ecoregions with data on biological distinctiveness offers a powerful tool for determining the relative conservation importance of different ecoregions within a major habitat type. In this chapter, we present the results of the integration exercise; the model and decision rules are described in Chapter 1.

Results

The classification of ecoregions as level I, II, III, or IV is presented visually in Map 8 and is summarized in Tables 5-1 and 5-2. Note that the red areas in Map 8 (level I) cover large areas that may contain very limited intact habitats. With the exception of some western Amazonian units, remaining natural habitats in most ecoregions classified as Highest Priority at Regional Scale are only a fraction of the area shown in red. Tables 5-3 to 5-12 present the results by MHT.

Following the rules presented in Chapter 1, 19 ecoregions were upgraded to the Highest Priority at Regional Scale ranking to ensure bioregional representation. We designate these ecoregions as level I[a] ecoregions (Map 9). In the following list, the former biodiversity conservation priority level and final conservation status are given in parentheses:[1]

- Tehuantepec moist forests (II, Endangered)
- Cuban dry forests (II, Endangered)
- Tamaulipas/Veracruz dry forests (III, Endangered)
- Llanos dry forests (III, Endangered)

- Bolivian montane dry forests (II, Critical)
- Brazilian *Araucaria* forests (II, Critical)
- Tabasco/Veracruz savannas (III, Critical)
- Pampas (II, Endangered)
- Beni savannas (II, Endangered)
- Jalisco palm savannas (II, Critical)
- Guayaquil flooded grasslands (II, Endangered)
- Eastern Amazonian flooded grasslands (III, Vulnerable)[2]
- Mexican alpine tundra (III, Vulnerable)
- Leeward Islands xeric scrub (III, Critical)
- Pueblan xeric scrub (II, Critical)
- Araya and París xeric scrub (II, Endangered)
- Sechura desert (III, Vulnerable)
- Caatinga (III, Vulnerable)
- Paraguaná restingas (II, Endangered)

Ecoregions of Highest Biodiversity Conservation Priority by MHT

Among the 55 Tropical moist broadleaf forest ecoregions, 23 are classified as Highest Priority at Regional Scale, numerically the most among any MHT (Tables 5-1 and 5-3). One additional level I[a] ecoregion was selected in Central America. Attention should be given to two prominent level II ecoregions, the Tepuis of the Orinoco bioregion and the Japura/Negro moist forests of Amazonia, which are classified as Globally Outstanding and Relatively Intact. *These are the only ecoregions in Latin America and the Caribbean with this combination of features.*

---

1. Either the Eastern Mexican grasslands or the Central Mexican grasslands should probably be upgraded as a representative of the grasslands/savannas/shrublands MHT in the Northern Mexico bioregion. Both are, however, unclassified because of lack of information.

2. According to the suggested rules for upgrading ecoregions, the São Luis flooded grasslands should have been chosen in place of the Eastern Amazonian flooded grasslands. We substitute the latter, however, because of some doubts as to whether or not the former area should even be designated as a distinct ecoregion.

Table 5-1. Conservation Importance of Ecoregions by Major Habitat Type

Major Habitat Type	*Number of Ecoregions in Each Conservation Priority Class*				
	Level I	Level II	Level III	Level IV	Unclassified
Tropical moist broadleaf forests	23	10	18	4	
Tropical dry broadleaf forests	5	6	18	2	
Temperate forests	2		1		
Tropical and subtropical coniferous forests	5	5	3	3	
Grasslands, savannas, and shrublands	2	2	8	2	2
Flooded grasslands	4	4	4	1	
Montane grasslands	8		3	1	
Mediterranean scrub	2				
Deserts and xeric shrublands	2	6	9	6	4
Restingas	2	1			
TOTAL	55	34	64	19	6

Tropical dry broadleaf forests contain only three Globally Outstanding ecoregions (Table 5-4). Among the 31 ecoregions, five are level I's, two of which occur in the Central American bioregion. To achieve better bioregional representation, four other ecoregions are upgraded to a Iᵃ status.

There are only three ecoregions in the temperate forest MHT, all in the Southern South America bioregion (Table 5-5). Two of the three are considered Highest Priority at Regional Scale. No further ecoregions are needed to achieve bioregional representation.

Tropical and subtropical coniferous forests contain two Globally Outstanding ecoregions, the Sierra Madre del Sur pine-oak forests and the Sierra Madre Occidental pine-oak forests (Table 5-6). Five of the 16 ecoregions are classified as level I, and all but one of the bioregions are represented in this category. Brazilian *Araucaria* forests are classed as Iᵃ to achieve bioregional representation. The Miskito pine forests are noteworthy as a relatively rare large unit of tropical lowland pine forest.

Grasslands, savannas, and shrublands contain one Globally Outstanding ecoregion, the Cerrado (Table 5-7). Among the 16 ecoregions, only two are Highest Priority at Regional Scale: the Cerrado and the Chaco savannas. To achieve better bioregional representation, three are added as Iᵃ ecoregions: Tabasco/Veracruz savannas, Beni savannas (Amazonia) and the Pampas. Two grassland ecoregions in Mexico are unclassified.

Flooded grasslands contain four ecoregions classed as Highest Priority at Regional Scale, includ-ing the Pantanal. To achieve better bioregional representation requires adding three other ecoregions (Table 5-8).

Montane grasslands contain 12 ecoregions, four of them Globally Outstanding (Table 5-9). All four are disjunct examples of paramo, a globally rare montane community that is most extensive in South America. Eight montane grassland ecoregions are Highest Priority at Regional Scale. The Mexican alpine tundra receives high consideration as a Iᵃ ecoregion.

Only two Mediterranean scrub ecoregions exist in LAC, the Californian coastal sage-chaparral and the Chilean matorral (Table 5-10). Both are considered Globally Outstanding and highly threatened and are therefore Highest Priority at Regional Scale.

Only two of the 27 desert and xeric shrubland ecoregions were selected as Highest Priority at Regional Scale: Northern Sonoran cactus scrub and the Galapagos Islands xerics (Table 5-11). We were unable to determine the conservation status of four ecoregions in this MHT; all are northern or central Mexican xerics. To achieve better bioregional representation, five additional ecoregions are selected as Iᵃ ecoregions.

There are only three ecoregions in the restingas MHT; both of the Brazilian representatives are Globally Outstanding and Critical and thus Highest Priority at Regional Scale (Table 5-12). Paraguaná restingas are added as a Iᵃ ecoregion to ensure bioregional representation.

To summarize, only 55 out of 178 ecoregions distributed among the 10 terrestrial MHTs were

Table 5-2. Final Conservation Status and Biological Distinctiveness of All Non-Mangrove Ecoregions

Biological Distinctiveness	Final Conservation Status					
	Critical (31)	Endangered (51)	Vulnerable (55)	Relatively Stable (27)	Relatively Intact (8)	Unclassified (6)
Globally Outstanding (34)	I (9)	I (6)	I (12)	I (5)	II (2)	
Regionally Outstanding (31)	I (2)	I (9)	I (14)	II (5)	III (2)	
Bioregionally Important (63)	II (7)	II (19)	III (19)	III (13)	IV (2)	(3)
Locally Important (50)	III (13)	III (17)	IV (11)	IV (4)	IV (2)	(3)

Note: The values in parentheses are the number of ecoregions in that category.

The roman numerals indicate biodiversity conservation priority classes:
Level I = Highest Priority at Regional Scale (shaded area)
Level II = High Priority at Regional Scale
Level III = Moderate Priority at Regional Scale
Level IV = Important at National Scale

designated as level I. Application of our method to achieve bioregional representation adds 19 additional ecoregions designated as level I[a].

Combining these 19 ecoregions with the ecoregions classed as Highest Priority at Regional Scale gives a total of 74 units, 42 percent of all of the non-mangrove ecoregions.

Major Trends

Only two ecoregions, the Tepuis of the Orinoco bioregion and the Japura/Negro moist forests of Amazonia, were classified as Relatively Intact and Globally Outstanding. The relative scarcity of Globally or Regionally Outstanding ecoregions classified as Relatively Stable or Relatively Intact is of regional concern (Table 5-2). These fourteen ecoregions include some Amazonian units that are extremely large, but represent only 8 percent of the total number of ecoregions analyzed. Conversely, the number of ecoregions that are either Globally or Regionally Outstanding and Critical (11) or Globally Outstanding and Endangered (6), account for only 10 percent of the total number of ecoregions.

Table 5–3. Tropical Moist Broadleaf Forests: Integration Matrix of Biological Distinctiveness and Conservation Status

Biological Distinctiveness	Final Conservation Status				
	Critical	*Endangered*	*Vulnerable*	*Relatively Stable*	*Relatively Intact*
Globally Outstanding	• **Cauca Valley montane forests**—Colombia (N.Andes) • **Magdalena Valley montane forests**—Colombia (N.Andes) • **Brazilian Coastal Atlantic forests**—Brazil (E.S.America)	• **Northwestern Andean montane forests**—Colombia, Ecuador (N.Andes) • **Venezuelan Andes montane forests**—Venezuela, Colombia (N.Andes) • **Peruvian Yungas**—Peru (C.Andes)	• **Ucayali moist forests**—Brazil, Peru (Amazonia) • **Varzea forests**—Brazil, Peru, Colombia (Amazonia) • **Choco/Darién moist forests**—Colombia, Panama (N.Andes) • **Cordillera Oriental montane forests**—Colombia, Venezuela (N.Andes) • **Eastern Cordillera Real montane forests**—Ecuador, Colombia, Peru (N.Andes)	• **Napo moist forests**—Peru, Ecuador, Colombia (Amazonia) • **Western Amazonian swamp forests**—Peru, Colombia (Amazonia) • **Southwestern Amazonian moist forests**—Brazil, Peru, Bolivia (Amazonia)	• **Tepuis**—Venezuela, Brazil, Guyana, Suriname, Colombia (Amazonia) • **Japura/Negro moist forests**—Colombia, Venezuela, Brazil, Peru (Amazonia)
Regionally Outstanding	• **Western Ecuador moist forests**—Ecuador, Colombia (N.Andes)	• **Hispaniolan moist forests**—Haiti, Dominican Republic (Carib.) • **Jamaican moist forests**—Jamaica (Carib.) • **Bolivian Yungas**—Bolivia, Argentina (C.Andes) • **Brazilian Interior Atlantic forests**—Brazil, Argentina, Paraguay. (E.S.America)	• **Cuban moist forests**—Cuba (Carib.) • **Cordillera La Costa montane forests**—Venezuela (Orinoco) • **Macarena montane forests**—Colombia (Amazonia) • **Rondônia/Mato Grosso moist forests**—Brazil, Bolivia (Amazonia) • **Santa Marta montane forests**—Colombia (N.Andes)	• **Talamancan montane forests**—Costa Rica, Panama (C.America)	• **Guianan Highlands moist forests**—Venezuela, Brazil, Guyana (Orinoco) • **Juruá moist forests**—Brazil (Amazonia)

(Table continues on the following page)

Table 5-3. Tropical Moist Broadleaf Forests: Integration Matrix of Biological Distinctiveness and Conservation Status (*Continued*)

Biological Distinctiveness	Final Conservation Status				
	Critical	*Endangered*	*Vulnerable*	*Relatively Stable*	*Relatively Intact*
Bioregionally Outstanding		• Oaxacan moist forests — Mexico (C.America) • Tehuantepec moist forests — Mexico, Guatemala, Belize (C.America) • Sierra Madre moist forests — Mexico, Guatemala, El Salvador (C.America) • Central American montane forests — Mexico, El Salvador, Guatemala, Honduras (C.America) • Isthmian–Pacific moist forests — Costa Rica, Panama (C.America) • Magdalena/Urabá moist forests — Colombia (N.Andes)	• Puerto Rican moist forests — Puerto Rico (Carib.) • Windward Islands moist forests — Windward Islands (Carib.) • Yucatán moist forests — Mexico (C.America) • Central American Atlantic moist forests — Guatemala, Belize, Honduras, Nicaragua, Costa Rica, Panama (C.America) • Paramaribo swamp forests — Suriname (Amazonia) • Eastern Panamanian montane forests — Panama, Colombia (N.Andes) • Andean Yungas — Argentina, Bolivia (C.Andes)	• Leeward Islands moist forests — Leeward Islands (Carib.) • Orinoco Delta swamp forests — Venezuela, Guyana (Orinoco) • Uatama moist forests — Brazil, Venezuela, Guyana (Amazonia) • Amapá moist forests — Brazil, Suriname (Amazonia) • Guianan moist forests — Venezuela, Guyana, Suriname, Brazil, French Guiana (Amazonia)	
Locally Important	• Costa Rican seasonal moist forests — Costa Rica, Nicaragua (C.America) • Catatumbo moist forests — Venezuela, Colombia (N.Andes)	• Belizean swamp forests — Belize (C.America) • Tocantins moist forests — Brazil (Amazonia)	• Trinidad & Tobago moist forests — Trinidad & Tobago (Orinoco) • Tapajós/Xingu moist forests — Brazil (Amazonia)	• Purus/Madeira moist forests — Brazil (Amazonia) • Beni swamp and gallery forests — Bolivia, Brazil (Amazonia)	

☐ Level I: Highest Priority at Regional Scale for biodiversity conservation

⬚ Level 1a: Highest Priority at Regional Scale (added to ensure bioregional representation)

Table 5-4. Tropical Dry Broadleaf Forests: Integration Matrix of Biological Distinctiveness and Conservation Status

Biological Distinctiveness	Final Conservation Status				
	Critical	Endangered	Vulnerable	Relatively Stable	Relatively Intact
Globally Outstanding	• **Bolivian lowland dry forests**—Bolivia (Amazonia) • **Ecuadorian dry forests**—Ecuador (N.Andes)	• **Tumbes/Piura dry forests**—Ecuador, Peru (N.Andes)			
Regionally Outstanding		• **Jalisco dry forests**—Mexico (C.America) • **Balsas dry forests**—Mexico (C.America)			
Bioregionally Outstanding	• **Central American Pacific dry forests**—Mexico, Guatemala, El Salvador, Honduras, Nicaragua, Costa Rica (C.America) • **Bolivian montane dry forests**—Bolivia (C.Andes)	• **Cuban dry forests**—Cuba (Carib.) • **Hispaniolan dry forests**—Haiti, Dominican Republic (Carib.) • **Oaxacan dry forests**—Mexico (C.America) • **Marañón dry forests**—Peru (N.Andes)	• **Sinaloan dry forests**—Mexico (N.Mexico)		
Locally Important	• **Leeward Islands dry forests**—Leeward Islands (Carib.) • **Veracruz dry forests**—Mexico (C.America) • **Panamanian dry forests**—Panama (C.America) • **Cauca Valley dry forests**—Colombia (N.Andes) • **Magdalena Valley dry forests**—Colombia (N.Andes) • **Patía Valley dry forests**—Colombia (N.Andes) • **Sinú Valley dry forests**—Colombia (N.Andes)	• **Jamaican dry forests**—Jamaica (Carib.) • **Puerto Rican dry forests**—Puerto Rico (Carib.) • **Cayman Islands dry forests**—Cayman Islands (Carib.) • **Windward Islands dry forests**—Windward Islands (Carib.) • **Tamaulipas/Veracruz dry forests**—Mexico (N.Mexico) • **Yucatán dry forests**—Mexico (C.America) • **Llanos dry forests**—Venezuela (Orinoco) • **Trinidad & Tobago dry forests**—Trinidad & Tobago (Orinoco) • **Maracaibo dry forests**—Venezuela (N.Andes) • **Lara/Falcón dry forests**—Venezuela (N.Andes)	• **Bahamian dry forests**—Bahamas, Turks & Caicos Islands (Carib.)	• **Baja California dry forests**—Mexico (N.Mexico)	

Level I: Highest Priority at Regional Scale for biodiversity conservation

Level Ia: Highest Priority at Regional Scale (added to ensure bioregional representation)

Table 5-5. Temperate Forests: Integration Matrix of Biological Distinctiveness and Conservation Status

Biological Distinctiveness	Final Conservation Status				
	Critical	Endangered	Vulnerable	Relatively Stable	Relatively Intact
Globally Outstanding			• Valdivian temperate forests—Chile, Argentina (S.S.America)		
Regionally Outstanding		• Chilean winter-rain forests—Chile (S.S.America)			
Bioregionally Outstanding			• Subpolar *Nothofagus* forests—Chile, Argentina (S.S.America)		
Locally Important					

Level I: Highest Priority at Regional Scale for biodiversity conservation

Table 5-6. Tropical and Subtropical Coniferous Forests: Integration Matrix of Biological Distinctiveness and Conservation Status

Biological Distinctiveness	Final Conservation Status				
	Critical	Endangered	Vulnerable	Relatively Stable	Relatively Intact
Globally Outstanding	• Sierra Madre del Sur pine–oak forests—Mexico (C.America)	• Sierra Madre Occidental pine–oak forests—Mexico, USA (N.Mexico)		• Sierra Madre Oriental pine–oak forests—Mexico (N.Mexico) • Belizean pine forests—Belize (C.America) • Miskito pine forests—Nicaragua, Honduras (C.America)	
Regionally Outstanding		• Mexican transvolcanic pine–oak forests—Mexico (C.America)	• Cuban pine forests—Cuba (Carib.)	• Veracruz montane forests—Mexico (C.America)	
Bioregionally Outstanding	• Brazilian Araucaria forests—Brazil, Argentina (E.S.America)	• Central Mexican pine–oak forests—Mexico (N.Mexico)	• Hispaniolan pine forests—Haiti, Dominican Republic (Carib.) • Central American pine–oak forests—Mexico, Guatemala, Honduras, El Salvador, Nicaragua (C.America)		
Locally Important	• Veracruz pine–oak forests—Mexico (N.Mexico)		• Bahamian pine forests—Bahamas, Turks & Caicos Islands (Carib.) • Sierra Juarez pine–oak forests—Mexico, USA (N.Mexico)		• San Lucan pine–oak forests—Mexico (N.Mexico)

Level I: Highest Priority at Regional Scale for biodiversity conservation

Level 1ª: Highest Priority at Regional Scale (added to ensure bioregional representation)

Table 5-7. Grasslands, Savannas, and Shrublands: Integration Matrix of Biological Distinctiveness and Conservation Status

Biological Distinctiveness	Final Conservation Status				
	Critical	Endangered	Vulnerable	Relatively Stable	Relatively Intact
Globally Outstanding			• Cerrado—Brazil, Paraguay, Bolivia (E.S.America)		
Regionally Outstanding			• Chaco savannas—Argentina, Paraguay, Bolivia, Brazil (E.S.America)		
Bioregionally Outstanding		• Beni savannas—Bolivia (Amazonia) • Pampas—Argentina (S.S.America)	• Argentine Espinal—Argentina (S.S.America) • Uruguayan savannas—Uruguay, Brazil, Argentina (S.S.America)	• Llanos—Venezuela, Colombia (Orinoco) • Guianan savannas—Suriname, Guyana, Venezuela, Brazil (Amazonia) • Amazonian savannas—Brazil, Colombia, Venezuela (Amazonia) • Argentine Monte—Argentina (S.S.America)	
Locally Important	• Tabasco/Veracruz savannas—Mexico (C.America) • Tehuantepec savannas—Mexico (C.America)		• Humid Chaco—Argentina, Paraguay, Bolivia, Brazil (E.S.America) • Córdoba montane savannas—Argentina (E.S.America)		

Note: The final conservation status of two ecoregions in this MHT are unclassified: Central Mexican grasslands (Bioregionally Outstanding) and Eastern Mexican grasslands (Locally Important).

[ ] Level I: Highest Priority at Regional Scale for biodiversity conservation

[ ] Level 1ᵃ: Highest Priority at Regional Scale (added to ensure bioregional representation)

Table 5-8. Flooded Grasslands: Integration Matrix of Biological Distinctiveness and Conservation Status

Biological Distinctiveness	Final Conservation Status				
	Critical	*Endangered*	*Vulnerable*	*Relatively Stable*	*Relatively Intact*
Globally Outstanding			• **Pantanal** — Brazil, Bolivia, Paraguay (E.S.America)		
Regionally Outstanding	• **Central Mexican wetlands** — Mexico (N.Mexico)	• **Cuban wetlands** — Cuba (Carib.)	• **Enriquillo wetlands** — Haiti, Dominican Republic (Carib.)		• **Quintana Roo wetlands** — Mexico (C.America)
Bioregionally Outstanding	• **Jalisco palm savannas** — Mexico (C.America) • **Veracruz palm savannas** — Mexico (C.America)	• **Guayaquil flooded grasslands** — Ecuador (N.Andes) • **Paraná flooded savannas** — Argentina (E.S.America)	• **Eastern Amazonian flooded grasslands** — Brazil (Amazonia)	• **Orinoco wetlands** — Venezuela (Orinoco) • **Western Amazonian flooded grasslands** — Peru, Bolivia (Amazonia)	
Locally Important		• **São Luis flooded grasslands** — Brazil (Amazonia)			

Level I: Highest Priority at Regional Scale for biodiversity conservation

Level I*: Highest Priority at Regional Scale (added to ensure bioregional representation)

Table 5-9. Montane Grasslands: Integration Matrix of Biological Distinctiveness and Conservation Status

Biological Distinctiveness	Final Conservation Status				
	Critical	Endangered	Vulnerable	Relatively Stable	Relatively Intact
Globally Outstanding			• **Santa Marta paramo**—Colombia (N.Andes) • **Northern Andean paramo**—Colombia, Ecuador (N.Andes) • **Cordillera Central paramo**—Ecuador, Peru (C.Andes)	• **Cordillera de Mérida paramo**—Venezuela (N.Andes)	
Regionally Outstanding			• **Central Andean puna**—Bolivia, Argentina, Peru, Chile (C.Andes) • **Central Andean wet puna**—Peru, Bolivia, Chile (C.Andes) • **Central Andean dry puna**—Argentina, Bolivia, Chile (C.Andes) • **Patagonian steppe**—Argentina, Chile (S.S.America)		
Bioregionally Outstanding			• **Mexican alpine tundra**—Mexico (C.America) • **Patagonian grasslands**—Argentina, Chile (S.S.America)	• Costa Rican paramo—Costa Rica (C.America)	
Locally Important				• **Southern Andean steppe**—Argentina, Chile (S.S.America)	

Level I: Highest Priority at Regional Scale for biodiversity conservation

Level 1ᵃ: Highest Priority at Regional Scale (added to ensure bioregional representation)

Table 5-10. Mediterranean Scrub: Integration Matrix of Biological Distinctiveness and Conservation Status

Biological Distinctiveness	Final Conservation Status				
	Critical	*Endangered*	*Vulnerable*	*Relatively Stable*	*Relatively Intact*
Globally Outstanding	• California coastal sage-chaparral—Mexico, USA (N.Mexico)	• Chilean matorral—Chile (C.Andes)			
Regionally Outstanding					
Bioregionally Outstanding					
Locally Important					

Level I: Highest Priority at Regional Scale for biodiversity conservation

Table 5-11. Deserts and Xeric Shrublands: Integration Matrix of Biological Distinctiveness and Conservation Status

Biological Distinctiveness	Final Conservation Status				
	Critical	Endangered	Vulnerable	Relatively Stable	Relatively Intact
Globally Outstanding			• Galapagos Islands xeric scrub—Ecuador (N.Andes)	• Northern Sonoran cactus scrub—Mexico, USA (N.Mexico)	
Regionally Outstanding				• Sonoran xeric scrub—Mexico, USA (N.Mexico)	
Bioregionally Outstanding	• Pueblan xeric scrub—Mexico (C.America) • Motagua Valley thornscrub—Guatemala (C.America)	• Araya & Pariá xeric scrub—Venezuela (Orinoco) • Guajira/Barranquilla xeric scrub—Colombia, Venezuela (N.Andes) • Paraguaná xeric scrub—Venezuela (N.Andes)	• Guerreran cactus scrub—Mexico (C.America) • Sechura desert—Ecuador, Chile, Peru (C.Andes) • Atacama desert—Chile (C.Andes) • Caatinga—Brazil (E.S.America)		• Baja California xeric scrub—Mexico (N.Mexico)
Locally Important	• Leeward Islands xeric scrub—Leeward Islands (Carib.)	• Cayman Islands xeric scrub—Cayman Islands (Carib.) • Windward Islands xeric scrub—Windward Islands (Carib.) • Central Mexican mezquital—Mexico (N.Mexico) • La Costa xeric shrublands—Venezuela (Orinoco)	• Cuban cactus scrub—Cuba (Carib.) • Chihuahuan xeric scrub—Mexico, USA (N.Mexico) • Eastern Mexican mezquital—Mexico, USA (N.Mexico) • Aruba/Curaçao/Bonaire cactus scrub—Netherlands Antilles (Orinoco)		• San Lucan mezquital—Mexico (N.Mexico)

Note: The final conservation status of four ecoregions in this MHT is unclassified: Western Mexican mezquital and Mexican Interior chaparral (both Locally Important) and Eastern Mexican matorral and Central Mexican cactus scrub (both Bioregionally Outstanding).

Level I: Highest Priority at Regional Scale for biodiversity conservation

Level 1a: Highest Priority at Regional Scale (added to ensure bioregional representation)

Table 5-12. Restingas: Integration Matrix of Biological Distinctiveness and Conservation Status

Biological Distinctiveness	Final Conservation Status				
	Critical	Endangered	Vulnerable	Relatively Stable	Relatively Intact
Globally Outstanding	• **Northeastern Brazil restingas** — Brazil (Amazonia) • **Brazilian Atlantic Coast restingas** — Brazil (E.S.America)				
Regionally Outstanding					
Bioregionally Outstanding		• **Paraguaná restingas** — Venezuela (N.Andes)			
Locally Important					

Level I: Highest Priority at Regional Scale for biodiversity conservation

Level 1ª: Highest Priority at Regional Scale (added to ensure bioregional representation)

6

Conservation Assessment of Mangrove Ecosystems

Among the five major ecosystem types (METs), mangroves probably receive the least attention from conservation donors and agencies. Perhaps the major reason that mangroves are underrepresented in conservation programs is that they are of superficially similar appearance across the entire region since mangroves, *as a forest type*, consist of only a few tree species. However, a *mangrove ecosystem* contains a wide diversity of aquatic and marine species and in some cases mangrove ecosystems can rival other habitat types in alpha diversity when both terrestrial and aquatic species are included in species lists. It is important not to confuse the definition of a mangrove forest with that of a mangrove ecosystem.

This chapter summarizes the results of a conservation assessment exercise targeted specifically at mangrove ecosystems and draws heavily from a synopsis prepared by Dr. Gilberto Cintrón (USFWS) and other contributors. The exercise was undertaken by the WWF Conservation Science Program in parallel with the WWF/World Bank study of LAC ecoregions and included major contributions from the leading experts on mangrove ecosystems in LAC (see the Acknowledgments section).

We begin by first defining a mangrove ecosystem and then briefly summarizing some important attributes necessary to better understand appropriate conservation approaches. We then delineate the mangroves of LAC into a series of smaller units that permits prioritizing investments as in the non-mangrove ecoregions. We then define and apply the method for the mangrove analysis, adapted by the workshop participants from the non-mangrove approach used elsewhere in this report (a more detailed description of the method can be found in Appendix B). We then present the results of an assessment of conservation status of mangrove ecoregions and a list of priorities for investments in restoration, biodiversity conservation, and sustainable use.

Definition

The word "mangrove" describes a salt-tolerant forest ecosystem that occupies sheltered tropical and subtropical coastal estuarine environments. The constituent plant species (also called mangroves) are not closely related, but they share morphological, physiological, and reproductive adaptations that allow survival in very saline, waterlogged, and reduced substrates. These substrates are often subject to rapid changes. Mangrove forest ecosystems are different from other forests in that they receive inputs of matter and energy from both land and sea. These inputs include fresh water, sediments and nutrients from land and tidal flushing, and saline intrusions from the sea. These inputs act as energy subsidies that increase productivity and help maintain ecosystem processes and high rates of organic matter fixation. The efficient allotment of organic production allows mangroves to develop complex forests in areas subject to active geomorphic processes and change (Thom 1967, 1984), climatic stresses like storms (Craighead and Gilbert 1962; Tabb and Jones 1962), hypersalinity (Cintrón et al. 1978), and even occasional frosts (Lugo and Patterson-Zucca 1977).

Ecological Attributes of Mangrove Ecosystems

The relatively high plant productivity and the active biological processes characteristic of mangrove ecosystems yield many goods and services of direct or indirect benefit. In the LAC region mangroves are exploited for their timber, fuelwood, and charcoal.

They are also known to be important for estuarine fisheries because of the contribution of detritus and dissolved organic carbon to estuarine food webs and the shelter their roots provide for juveniles (Heald 1969; Odum 1971; Twilley 1982). Often unrecognized is the fact that mangrove ecosystems support organic matter fixation by other primary producers, and that the importance of mangroves also resides in the role they play in the maintenance of the geomorphic structure of these environments (e.g., estuaries, lagoons, and reefs).

Mangroves grow over landforms created and shaped by local geomorphic processes, creating a complex woody structure (trunks, branches, aerial roots, pneumatophores) that varies in degree of development and architecture as a response to physiographic and climatic conditions. The heterogeneity of landforms and forest architecture gives rise to a variety of habitats that provide shelter, foraging grounds, and nursery areas to many marine and terrestrial animals.

The production of large amounts of litterfall leads to large exports of particulate and dissolved organic carbon. High levels of secondary productivity are supported by the production of phytoplankton and benthic autotrophs (sea grasses and algae), and by detritus inputs from upland sources. This high productivity allows the establishment of a complex and diverse food web that may support large resident and migratory populations of mammals, reptiles, birds, fish, crustaceans, mollusks, and other associated animals.

The relative importance of various ecological processes in a given mangrove ecosystem is shaped by the geomorphic, hydrologic, and climatic characteristics of the area. Mangroves are also strongly coupled to adjacent coastal and terrestrial areas through ecologically important processes such as the movement of migratory species of fish and shrimp to coastal shelves and through bird and mammal populations moving between mangroves and inland areas. There are also ecological linkages, in terms of goods and services, binding mangroves, seagrass beds, and coral reefs—where these systems coexist.

Mangrove ecosystems perform other services as well. They play an important role in water storage and the trapping of sediments and carbon, contributing to the control of the quality and quantity of water, particulates, and solutes discharged to the ocean. The intricate network of roots that binds the substrate, the trunks and the root breathing organs (pneumatophores) dissipate water energy and promote the deposition of materials and reduces erosion. The underground root network contributes to bank and sediment stabilization. The degree of control which mangroves exert on nutrient and sediment fluxes and shoreline erosion can determine the extent of the dominance of mud or coral reef environments within a landscape. Extensive coastal erosion has occurred in some areas where mangroves have been removed, thus indirectly influencing coastal biological communities.

Delineation of Mangrove Complexes and Units

Mangrove forests vary greatly in structural and functional characteristics (Pool et al. 1977; Lugo and Snedaker 1974; Cintrón et al. 1985). The estuarine distribution of mangroves also gives them a rather linear appearance in contrast to other ecoregions. However, at the landscape level, geomorphic processes provide a range of discrete landform types over which mangroves become established (Hayes 1975; Galloway 1975; Thom 1984). For this study, mangroves were categorized on the basis of four major landscape types colonized by mangroves: deltas, estuaries, lagoons, and carbonate platforms (adapted from eight major types recognized by Thom 1967, 1984).

In order to begin delineating mangroves, we made the assumption that if a coast is divided into segments having comparable environmental conditions and physiography, each unit would be characterized by distinctive landforms (see above) and processes. Mangroves within each segment would (a) occupy the suite of suitable landform types typical of that energy regime; (b) develop ecosystems with similar attributes and outputs (levels of development and productivity); (c) exhibit similar vulnerabilities and responses to disturbance regimes; and (d) be most responsive to a similar set of conservation activities. This approach was used by Schaeffer-Novelli et al. (1990) to characterize the mangrove environments of the Brazilian coast. We adopted this concept since it provides a bioregional view with great potential for the development of conservation priorities and management strategies for mangroves.

Thirteen major biogeographic units, designated as complexes, were identified along the coastlines of Latin America and the Caribbean (Map 4; Table 6-1). Each complex was further subdivided to give a total of 40 smaller units or segments. Individual sites, corresponding to major landscape types, can be recognized within each segment. In this study we only present results at the complex and unit level. A detailed report on the results of the workshop, including descriptions of high priority sites and updated maps of mangrove localities, will be available in 1995 from WWF.

Table 6-1. Conservation Status of Mangrove Units of LAC

Mangrove Complex	Mangrove Unit	Snapshot Conservation Status	Threat	Final Conservation Status
Atlantic				
Gulf of Mexico	Alvarado	Vulnerable	Medium	Vulnerable
	Usumacinta	Relatively Stable	Medium	Relatively Stable
Yucatán	Petenes	Relatively Stable	Low	Relatively Stable
	Río Lagartos	Relatively Stable	Low	Relatively Stable
	Mayan Corridor	Vulnerable	High	Endangered
	Belizean Coast	Vulnerable	Medium	Vulnerable
	Belizean Reef	Vulnerable	Medium	Vulnerable
Atlantic Central America	Northern Honduras	Vulnerable	High	Vulnerable
	Mosquitia/Nicaraguan Caribbean Coast	Relatively Stable	Low	Relatively Stable
	Río Negro/Río San Sun	Vulnerable	Medium	Endangered
	Bocas del Toro/Bastimentos Island/San Blas	Relatively Stable	Medium	Vulnerable
West Indies	Bahamas	Vulnerable	Medium	Vulnerable
	Greater Antilles	Endangered	Medium	Endangered
	Lesser Antilles	Endangered	High	Critical
Continental Caribbean	Coastal Venezuela	Vulnerable	Medium	Vulnerable
	Magdalena/Santa Marta	Endangered	High	Endangered
Amazon-Orinoco-Maranhão	Trinidad	Vulnerable	High	Endangered
	Segment 0: Orinoco-Cabo Orange	Relatively Stable	Medium	Relatively Stable
	Segment I: Cabo Orange-Cabo Norte	Relatively Stable	Medium	Relatively Stable
	Segment II: Cabo Norte-Ponta Curuçá	Relatively Stable	Medium	Relatively Stable
	Segment III: Ponta Curuçá-Parnaíba Delta	Relatively Stable	Medium	Vulnerable
NE Brazil	Segment IV: Parnaíba Delta-Cabo Calcanhar	Vulnerable	Medium	Vulnerable
	Segment V: Cabo Calcanhar-Recôncavo Baiano	Vulnerable	Medium	Endangered
	Segment VI: Recôncavo Baiano-Cabo Frio	Vulnerable	Medium	Endangered
SE Brazil	Segment VII: Cabo Frio-Laguna	Vulnerable	Medium	Endangered

Mangrove Complex	Mangrove Unit	Snapshot Conservation Status	Threat	Final Conservation Status
Pacific				
Sea of Cortez	NW Mexican Coast	Vulnerable	High	Endangered
	Marismas Nacionales/ San Blas	Endangered	High	Endangered
Southern Mexico	Southern Pacific Coast of Mexico	Vulnerable	Medium	Vulnerable
Pacific Central America	Tehuantepec/El Manchón	Relatively Stable	Low	Relatively Stable
	Northern Dry Pacific Coast	Endangered	High	Endangered
	Gulf of Fonseca	Vulnerable	High	Endangered
	Southern Dry Pacific Coast	Endangered	High	Critical
	Moist Pacific Coast	Vulnerable	Medium	Vulnerable
	Panama Dry Pacific Coast	Relatively Stable	Low	Relatively Stable
Pacific South America	Gulf of Panama	Relatively Stable	Medium	Relatively Stable
	Esmeraldas/Pacific Colombia	Relatively Stable	Medium	Relatively Stable
	Manabí	Critical	High	Critical
	Gulf of Guayaquil/Tumbes	Vulnerable	High	Endangered
	Piura	Relatively Stable	High	Vulnerable
Galapagos	Galapagos Islands	Unclassified	Unclassified	Unclassified

Conservation Status

We estimated the conservation status of mangrove units through assessments of the following criteria: loss of habitat; large blocks of intact habitat; water quality and hydrographic integrity; rate of habitat conversion; degree of protection; degree of fragmentation; and the degree of alteration of the catchment basin (refer to Appendix B for a more detailed discussion of the methods). As was done for the terrestrial ecoregions, the snapshot conservation status of the mangrove units was modified by assessments of current and projected threats (e.g., clearing for development, shrimp farming, and agriculture, pollution, wildlife exploitation and overharvesting of fisheries, and alteration of hydrologic flows through road building or channelization; Table 6-1).

Mangrove units were Relatively Stable in only five areas (13 percent): portions of the Yucatán Peninsula; the Pacific mangroves of Panama; the Mosquitia region of Central America; the Tehuantepec/

El Manchón region of Pacific Central America; and the South American Atlantic coast from just south of the Amazon Delta to the Orinoco Delta. No Relatively Intact mangrove units were recognized. Fifteen (38 percent) mangrove units were either Critical or Endangered.

Biological Distinctiveness and Conservation Activities

Although the workshop identified individual mangrove sites that had outstanding ecological or biological features (e.g., unusual species assemblages, critical habitat for migratory birds), no attempt was made to rank mangrove units on the basis of their biological distinctiveness because (a) discrimination among mangroves on the basis of community diversity is difficult as mangroves throughout the LAC region share a high proportion of species and have relatively similar high levels of alpha diversity and low endemism; and (b) mangrove ecosystems are

Table 6-2. Most Appropriate Conservation Activities for Mangrove Units of LAC

Mangrove Complex	Restoration	Conservation with Restricted Access	Conservation for Sustainable Use
Gulf of Mexico		Usumacinta 1	Alvarado 3; Usamacinta 1
Yucatán		Belizean Reef 5; Belizean Coast 1; Río Lagartos 3	Mayan Corridor 4; Belizean Coast 2; Petenes 3
Atlantic Central America		Mosquitia/Nicaraguan Caribbean Coast 1; Northern Honduras 3	Northern Honduras 3; Río Negro/Río San Sun 3; Bocas del Toro/Bastimentos Island/San Blas 4
West Indies	Lesser Antilles 2	Greater Antilles (Cuba only) 2; Lesser Antilles (Guadalupe) 2	Bahamas 5; Greater Antilles (excluding Cuba) 2
Continental Caribbean	Magdalena/ Santa Marta 2	Coastal Venezuela 1; Magdalena/Santa Marta 4	Coastal Venezuela 3
Amazon-Orinoco-Maranhão		Orinoco 1; Segment 0 1; Segment III 1	Trinidad 2; Segment III 1; Segment I 1; Segment II 1
NE Brazil	Segment IV 5; Segment V 4; Segment VI 3		Segment V 2; Segment VI 2
SE Brazil	Segment VII 4		Segment VII 4
Sea of Cortez	Marismas Nacionales/ San Blas 3	Marismas Nacionales/San Blas 1	NW Mexican Coast 5; Marismas Nacionales/San Blas 1
Southern Mexico			Southern Pacific Coast of Mexico 4
Pacific Central America	Southern Dry Pacific Coast 1	Moist Pacific Coast (between Panama and Costa Rica) 2; Panama Dry Pacific Coast 4; Tehuantepec/El Manchón 1	Moist Pacific Coast (between Panama and Costa Rica) 1; Panama Dry Pacific Coast 3; Gulf of Fonseca 2; Northern Dry Pacific Coast 3; Tehuantepec/ El Manchón 2
Pacific South America	Manabí 1	Esmeraldas/Pacific Colombia 3; Piura 3; Gulf of Panama 2	Esmeraldas/Pacific Colombia (southern part) 2; Gulf of Guayaquil/Tumbes 1; Gulf of Panama 2
Galapagos		Galapagos Islands 3	
Total Units	**9**	**21**	**29**

Note: Mangrove units are placed in the conservation activity category deemed most appropriate (some are placed in more than one activity column) according to the following criteria: (a) Restoration (ecological feasibility of restoration; relative beneficial ecological impact from restoration); (b) Conservation with restricted access (distinctiveness of biota and wildlife value; relative importance of ecosystem services assured through strict protection); and (c) Conservation for sustainable use (diversity of resources; capacity of system for recovery or resiliency). Each unit receives a priority ranking (1 being the highest and 5 the lowest) reflecting the urgency with which the unit should be considered.

exceptionally important in maintaining populations and ecological processes in surrounding marine, freshwater, and terrestrial ecosystems. Thus, all mangroves were recognized as "keystone" ecosystems, whose persistence is critical for the functioning of diverse and extensive ecosystems well beyond the boundaries of an individual mangrove forest.

Because of the important ecological role of mangroves, the workshop team stressed that the conservation of all mangroves should be a priority and chose to categorize mangrove units on the basis of the conservation activity most appropriate for each: restoration; conservation with restricted access (e.g., protected areas); and conservation with an emphasis on sustainable use (Table 6-2). Within each category, mangrove units were ranked on the basis of how important and feasible each of the assigned activities would be for the habitat and surrounding ecosystems. For example, in the restoration category, mangrove units were ranked according to the relative ecological impact and ecological feasibility of successful restoration.

For conservation with restricted access, different units were assessed in terms of the distinctiveness of their associated biota and value for wildlife, and the level of ecosystem services that would be assured through strict protection. The diversity of resources

and the capacity for recovery were used to rank mangrove units within the conservation for sustainable use category. These analyses are intended to help guide donors in choosing appropriate conservation activities for each mangrove unit and to provide a preliminary ecological cost-benefit analysis for the units within each activity category. These rankings should guide the spatial and temporal sequence of conservation activities, and not be used to justify unsustainable uses of particular mangrove systems.

Restoration was identified as the most appropriate conservation activity for nine mangrove units, conservation with restricted access for twenty units, and conservation with sustainable use for twenty-nine units. Seventeen units were categorized under two or more activities.

The results of the workshop are presently under final review by mangrove experts and the updated maps are being input into a geographic information system (GIS). We look forward to integrating the results of the mangrove workshop with those for the terrestrial ecoregions and upcoming analyses of freshwater and marine ecosystems. Ecosystem dynamics require that linkages among terrestrial, freshwater, mangrove, and marine systems are increasingly emphasized if we are to develop effective priority-setting methods and conservation strategies.

7

Conclusions and Recommendations

It is important to conserve biodiversity in every ecoregion. This study is not intended to diminish or discourage conservation programs in any ecoregion, particularly ecoregions identified as lower priority, but rather to assist conservation donors in planning the timing, sequence, and level of funding of current and future conservation efforts.

This study identifies the ecoregions of highest biodiversity conservation priority based on final conservation status and biological distinctiveness. All high-priority ecoregions merit immediate attention from donors. Ultimately, investment priorities can be derived only after passing the list of priority sites through other political, social, institutional, and economic filters. We feel that these filters are most accurate and useful at the intra-ecoregion scale rather than in comparisons among ecoregions.

In this chapter we review the results of the integration model and make several broad recommendations. We also discuss the results with respect to several other efforts to use biodiversity indicators to set regional priorities and explore the next steps needed to refine biodiversity priorities into investment priorities. Finally, we examine the application of this method to select priority sites at subregional and national scales.

The integration exercise identifies 55 ecoregions as being of Highest Priority at Regional Scale (level I). The distribution of these ecoregions by bioregion shows a relatively wide scatter (Table 7-1). The table also includes those 19 ecoregions of elevated rank added to achieve better bioregional representation (level Iª). The Northern Andes bioregion has the greatest number of ecoregions classed Highest Priority at Regional Scale. With the exception of Northern Mexico, Orinoco, and Southern South America, ecoregions of Highest Priority at Regional Scale occur in roughly equal numbers in the other six bioregions. Table 7-1 could serve as a starting point

for major donors interested in investing in biodiversity conservation in the region. The list and the integration exercises for each MHT (Tables 5-3 to 5-12) can also help donors to:

- Intervene quickly to ward off complete degradation and conversion in Critical ecoregions, particularly where threats to biodiversity are projected to remain intense over the next five years
- Identify ecoregions that, due to their conservation status, cannot absorb further intensive development projects
- Flag those high beta diversity ecoregions that run a high risk of species extinctions if threatened by extensive habitat conversion associated with major development projects
- Identify ecoregions best suited for environmentally sound sustainable development projects: those ecoregions that are largely intact, those with average (i.e., locally important) biodiversity, and those falling in an MHT characterized by low beta diversity
- Conduct a much needed overlay analysis of the allocation of donor funding by ecoregion, within a MHT or bioregion.

Seventeen ecoregions are classified as (a) Globally Outstanding and either Critical or Endangered, or (b) Regionally Outstanding and Critical (the three cells in the upper left corner of the integration matrix shown in Table 1-1). These ecoregions (shown in italics on Table 7-1) require immediate attention because of (a) the extraordinary or very unusual types of biodiversity found within them, and (b) the possibility that many of the remaining habitat blocks in these ecoregions will be completely degraded within 5 to 20 years, resulting in a loss of many species endemic to these units, particularly plants and invertebrates.

Table 7-1. Ecoregions of Highest Priority at Regional Scale by Bioregion and Major Habitat Type

Bioregion (Countries partially or entirely within the bioregion)	Major habitat type Level I = Highest Priority at Regional Scale *Level I (in italics) = Critical and Globally/Regionally Outstanding or Endangered and Globally Outstanding* Level Iᵃ = Ecoregion considered of Highest Priority at Regional Scale to achieve bioregional representation (all of these were originally classified as level II or III)
Caribbean (Bahamas, Greater Antilles, Lesser Antilles, Netherlands Antilles, Turks and Caicos Islands, Cayman Islands)	**Tropical Moist Broadleaf Forests** I–Cuban moist forests–Cuba I–Hispaniolan moist forests–Haiti, Dominican Republic I–Jamaican moist forests–Jamaica **Tropical Dry Broadleaf Forests** Iᵃ–Cuban dry forests–Cuba **Tropical and Subtropical Coniferous Forests** I–Cuban pine forests–Cuba I–Hispaniolan pine forests–Haiti, Dominican Republic **Flooded Grasslands** I–Cuban wetlands–Cuba I–Enriquillo wetlands–Haiti, Dominican Republic **Deserts and Xeric Shrublands** Iᵃ–Leeward Islands xeric scrub–Leeward Islands
Northern Mexico (Mexico, USA)	**Tropical Dry Broadleaf Forests** Iᵃ–Tamaulipas/Veracruz dry forests–Mexico **Tropical and Subtropical Coniferous Forests** *I–Sierra Madre Occidental pine-oak forests–Mexico, USA* **Flooded Grasslands** *I–Central Mexican wetlands–Mexico* **Mediterranean Scrub** *I–California coastal sage-chaparral–Mexico, USA* **Deserts and Xeric Shrublands** I–Northern Sonoran cactus scrub–Mexico, USA
Central America (Mexico, Belize, Guatemala, Honduras, Nicaragua, Costa Rica, Panama, El Salvador)	**Tropical Moist Broadleaf Forests** Iᵃ–Tehuantepec moist forests–Mexico, Guatemala, Belize **Tropical Dry Broadleaf Forests** I–Jalisco dry forests–Mexico I–Balsas dry forests–Mexico **Tropical and Subtropical Coniferous Forests** I–Mexican transvolcanic pine-oak forests–Mexico *I–Sierra Madre del Sur pine-oak forests–Mexico* **Grasslands, Savannas, and Shrublands** Iᵃ–Tabasco/Veracruz savannas–Mexico **Flooded Grasslands** Iᵃ–Jalisco palm savannas–Mexico **Montane Grasslands** Iᵃ–Mexican alpine tundra–Mexico **Deserts and Xeric Shrublands** Iᵃ–Pueblan xeric scrub–Mexico

(Table continues on the following page)

Table 7-1. Ecoregions of Highest Priority at Regional Scale by Bioregion and Major Habitat Type *(continued)*

Bioregion (Countries partially or entirely within the bioregion)	Major habitat type Level I = Highest Priority at Regional Scale *Level I (in italics) = Critical and Globally/Regionally Outstanding or Endangered and Globally Outstanding* Level Iᵃ = Ecoregion considered of Highest Priority at Regional Scale to achieve bioregional representation (all of these were originally classified as level II or III)
Orinoco (Trinidad and Tobago, Venezuela, Colombia, Brazil, Guyana)	**Tropical Moist Broadleaf Forests** I–Cordillera La Costa montane forests — Venezuela **Tropical Dry Broadleaf Forests** Iᵃ–Llanos dry forests — Venezuela **Deserts and Xeric Shrublands** Iᵃ–Araya and Paría xeric scrub — Venezuela
Amazonia (Brazil, Colombia, Venezuela, Guyana, French Guiana, Suriname, Bolivia, Peru, Ecuador)	**Tropical Moist Broadleaf Forests** I–Napo moist forests — Peru, Ecuador, Colombia I–Macarena montane forests — Colombia I–Ucayali moist forests — Brazil, Peru I–Western Amazonian swamp forests — Peru, Colombia I–Southwestern Amazonian moist forests — Brazil, Peru, Bolivia I–Varzea forests — Brazil, Peru, Colombia **Tropical Dry Broadleaf Forests** *I–Bolivian lowland dry forests — Bolivia, Brazil* **Grasslands, Savannas, and Shrublands** Iᵃ–Beni savannas — Bolivia **Flooded Grasslands** Iᵃ–Eastern Amazonian flooded grasslands — Brazil
Northern Andes (Venezuela, Colombia, Panama, Ecuador, Peru)	**Tropical Moist Broadleaf Forests** I–Chocó/Darién moist forests — Colombia, Panama *I–Northwestern Andean montane forests — Colombia, Ecuador* *I–Western Ecuador moist forests — Ecuador, Colombia* *I–Cauca Valley montane forests — Colombia* *I–Magdalena Valley montane forests — Colombia* I–Cordillera Oriental montane forests — Colombia, Venezuela I–Eastern Cordillera Real montane forests — Ecuador, Colombia, Peru I–Santa Marta montane forests — Colombia *I–Venezuelan Andes montane forests — Venezuela, Colombia* **Tropical Dry Broadleaf Forests** *I–Ecuadorian dry forests — Ecuador* *I–Tumbes/Piura dry forests — Ecuador, Peru* **Flooded Grasslands** Iᵃ–Guayaquil flooded grasslands — Ecuador **Montane Grasslands** I–Santa Marta paramo — Colombia I–Cordillera de Mérida paramo — Venezuela I–Northern Andean paramo — Colombia, Ecuador **Deserts and Xeric Shrublands** I–Galapagos Islands xeric scrub — Ecuador **Restingas** Iᵃ–Paraguaná restingas — Venezuela

Table 7-1. Ecoregions of Highest Priority at Regional Scale by Bioregion and Major Habitat Type *(continued)*

Bioregion (Countries partially or entirely within the bioregion)	Major habitat type Level I = Highest Priority at Regional Scale *Level I (in italics) = Critical and Globally/Regionally Outstanding or Endangered and Globally Outstanding* Level Iª = Ecoregion considered of Highest Priority at Regional Scale to achieve bioregional representation (all of these were originally classified as level II or III)
Central Andes (Peru, Bolivia, Chile, Argentina)	**Tropical Moist Broadleaf Forests** *I–Peruvian Yungas – Peru* I–Bolivian Yungas – Bolivia, Argentina **Tropical Dry Broadleaf Forests** Iª–Bolivian montane dry forests – Bolivia **Montane Grasslands** I–Cordillera Central paramo – Ecuador, Peru I–Central Andean puna – Bolivia, Argentina, Peru, Chile I–Central Andean wet puna – Peru, Bolivia, Chile I–Central Andean dry puna – Argentina, Bolivia, Chile **Mediterranean Scrub** *I–Chilean matorral – Chile* **Deserts and Xeric Shrublands** Iª–Sechura desert – Peru, Chile
Eastern South America (Brazil, Bolivia, Paraguay, Argentina)	**Tropical Moist Broadleaf Forests** *I–Brazilian Coastal Atlantic forests – Brazil* I–Brazilian Interior Atlantic forests – Brazil, Argentina, Paraguay **Tropical and Subtropical Coniferous Forests** Iª–Brazilian *Araucaria* forests – Brazil, Argentina **Grasslands, Savannas, and Shrublands** I–Cerrado – Brazil, Paraguay, Bolivia I–Chaco savannas – Argentina, Paraguay, Bolivia, Brazil **Flooded Grasslands** I–Pantanal – Brazil, Bolivia, Paraguay **Deserts and Xeric Shrublands** Iª–Caatinga – Brazil **Restingas** *I–Northeastern Brazil restingas – Brazil* *I–Brazilian Atlantic Coast restingas – Brazil*
Southern South America (Brazil, Uruguay, Argentina, Chile)	**Temperate Forests** I–Chilean winter-rain forests – Chile I–Valdivian forests – Chile, Argentina **Grasslands, Savannas, and Shrublands** Iª–Pampas – Argentina **Montane Grasslands** I–Patagonian steppe – Argentina, Chile

At the same time, the few ecoregions that contain relatively intact ecosystems deserve immediate conservation investment because ecosystem processes and species have the best chance for long-term persistence within these rare intact landscapes. It would be unwise to miss the window of opportunity they offer for strategic conservation planning. Some of the investments in these ecoregions may be limited to monitoring environmental conditions, lobbying against poorly conceived development projects, or other low-cost activities. At a minimum, planning and establishment of effective protected area systems should be initiated within each of these important areas.[1]

The purpose of this study is to provide a regional overview and an input to national biodiversity strategies. We offer here a few observations of interest to conservation planners at national levels. Among the countries of LAC, Ecuador is notable for being entirely covered by ecoregions of Highest Priority at Regional Scale (i.e., red or purple on Map 9). Peru, Bolivia, Cuba, Haiti, Dominican Republic, and Jamaica also are largely covered by such ecoregions, and Colombia, Chile, southern Brazil, and southern Mexico are not far behind.

Comparisons with Other Priority-Setting Frameworks for LAC

One of the more interesting opportunities offered by this study is to overlay results from other regional and subregional priority-setting exercises. The three most comprehensive efforts to date are the priority-setting exercise coordinated by the Biodiversity Support Program on behalf of USAID (BSP/CI/TNC/WRI/WWF 1995), an analysis of endemic bird areas in LAC by Birdlife International (Long et al. 1994; Wege and Long in press), and the database on Centers of Plant Diversity (WWF and IUCN 1994).

A number of the data layers and analyses conducted during the BSP exercise are also part of this study. Not surprisingly, there is considerable overlap in the results. The hierarchical classification scheme outlined in Chapter 1 served as the template for maintaining representation in the BSP exercise, and their map of "regional habitat units" was derived, in part, from the ecoregions map. The method to assess conservation status was virtually the same. The major differences in method from the BSP study include:

- Conducting the analysis at a finer geographic scale by using ecoregions rather than the more coarse-grained regional habitat units.
- Using additional criteria to achieve representation among MHTs and bioregional representation within MHTs.
- Linking biological distinctiveness to a biogeographic spatial scale and using four categories of biological distinctiveness rather than the three classes used in the BSP workshop.

Both studies identified as highest priority the Atlantic moist forests and the Cerrado of Brazil, the Patagonian steppes, the submontane and montane regions of the northern Andes, the montane grasslands of the central and northern Andes, the montane mixed conifer forests of Mexico, and the Sonoran desert ecoregion.

There were also some notable differences. This study recognized as Highest Priority at Regional Scale the Valdivian temperate forest, the Chilean winter-rain habitats (forest and scrub), the drier portion of the Chaco savannas, ecoregions composing the western arc of Amazonia, varzea moist forests of Amazonia, the Chocó/Darién moist forests, the moist forests of the Greater Antilles, and the dry forests of southern Mexico. These regions were considered of secondary or lower priority by the BSP exercise. In contrast, the BSP exercise classified many ecoregions among the Mexican xeric formations and the Beni savannas as highest priorities.

The results of the Birdlife and Centers of Plant Diversity studies were of value to us in the delimitation of ecoregions. It would be very instructive to closely compare the results of these two studies with the high-priority ecoregions we have identified. These two studies will also be useful for setting priorities at finer geographic scales (see below).

Application of the Methodology to Finer Geographic Scales

The integration of biological distinctiveness and conservation status helps target the highest priority ecoregions for urgent attention by donors interested in biodiversity conservation. By giving increased funding to these ecoregions, donors and governments will increase the likelihood of maintaining representation of all ecosystem and habitat types in a regional investment portfolio.

However, regional priority-setting exercises—by virtue of their coarse scale—can seldom identify where the most important investments should be made *within ecoregions* nor *what to do at those sites to best conserve biodiversity*. Prioritizing among 191 ecoregions at a regional scale is difficult enough.

---

1. Some ecoregions conceivably may be of high priority but stable with negligible conversion rate and good protection of remaining habitat. Since these factors tend to vary widely across an ecoregion, they should be evaluated at finer geographic scales.

Prioritizing among the tens, hundreds, or thousands of blocks of remaining habitat within an ecoregion is even more difficult. Yet, it is often at the site level that decisions must be made for conservation action.

Thus, the development of frameworks at finer geographic scales, such as intra-ecoregional analyses, are an essential second step of the priority-setting process. Without this framework, donors risk financing biodiversity conservation in the most important ecoregions, but conserving some of the less important, unique, or intact habitat blocks within them.

For example, even Globally Outstanding ecoregions contain some blocks of remaining habitat that are of relatively low biological or persistence value compared with other blocks in the same unit. Alternatively, even ecoregions designated as Locally Important may contain a few blocks larger and more intact than those found in other ecoregions within the same MHT and so would qualify as sites of high conservation interest. A framework that also interprets spatial patterns of biodiversity within an ecoregion can also help to determine the level of effort (e.g., number and size of protected areas) needed to conserve the biodiversity present.

As illustrated by this study, a biologically based framework can help to set priorities among ecoregions within an MHT. The same landscape ecology and conservation biology principles can help guide donors to the most biologically important blocks or clusters of habitats within ecoregions with the highest probabilities of persistence.

Several methods for determining priority sites for conservation, specifically protected areas, have been created (e.g., Margules et al. 1988; Pressey et al. 1993). All of these methods, however, demand extensive data—specifically, detailed information on the distribution of species representing many different taxa. Such data sets are largely incomplete for most ecoregions in LAC. They also pay less attention to landscape features that will influence the effectiveness of conservation of unique taxa at those sites.

To address this need, WWF is developing a method to set priorities among habitat blocks within ecoregions. The method consists of a series of analyses, the results of which can be overlaid using a GIS to determine intra-ecoregion conservation priorities. Briefly, four core analyses are needed to prepare overlays:

- *A persistence value analysis* of remaining habitat blocks or clusters of blocks to determine their relative probability of maintaining biodiversity over the long term. We use landscape features such as minimum size of blocks to conserve biodiversity and processes, configuration of remaining blocks, and analysis of intervening habitats to assess landscape integrity.
- *A biodiversity and critical habitats analysis* to overlay patterns of species richness hotspots, areas of high endemism or beta diversity, rare habitat types, populations of rare species, critical habitat for migrants and resident species, and core habitats within an ecoregion.
- *An analysis of existing protected areas and corridors* overlaid on the first two layers. This step helps to determine where investments should go first and which areas should receive proportionately greater attention.
- *A land use analysis* to determine the feasibility of creating new reserves and managing habitats adjacent to protected areas in a manner more compatible with conservation of biodiversity. This step assesses other uses or demands for resources by stakeholders who have access to or heavily utilize blocks with high conservation potential.

Integration of these layers can also identify priority areas that may never receive formal protection or were not classified as intact natural habitats. Because of their biodiversity value or their potential for enhancing the persistence of adjacent natural habitats, these areas may require restoration or management plans that emphasize conservation of biodiversity and natural resources.

These overlays incorporate the fundamental biological data that should serve as the foundation for setting investment priorities within ecoregions. After these overlays are performed, non-biological data can be used to further refine the list of options for investing in biodiversity. The methods described above can be used for all terrestrial ecoregions. By applying landscape-level criteria to their own maps of remaining habitat, as planners in Argentina are now doing, conservationists in LAC can better identify and prioritize among (a) the large blocks of remaining natural habitat with high potential for long-term biodiversity conservation, and (b) the biologically rich but highly fragmented habitats requiring urgent attention.

Appendix A

Methods Used for Assessing the Conservation Status of Terrestrial Ecoregions

An assessment of the conservation status of each ecoregion is intended to (a) identify major habitat types and ecoregions that are most threatened and thus help prioritize interventions to prevent their complete degradation or conversion; (b) create programs to conserve the most intact examples of major habitat types; and (c) help define appropriate conservation activities for different landscape scenarios. Here we define the conservation status of ecoregions in the tradition of the IUCN Red Data Book categories for threatened and endangered species. The conservation status categories we use for ecoregions are Extinct, Critical, Endangered, Vulnerable, Relatively Stable, and Relatively Intact. These categories represent different degrees of alteration and of spatial patterns of remaining habitats across landscapes. They reflect how with increasing habitat loss, degradation, and fragmentation, ecological processes cease to function naturally, or at all, and major components of biodiversity are steadily eroded. Landscape features are used as indicators for the ecological integrity of ecosystems.

Comments on the Approach

The conservation status assessment used here evolved out of an earlier approach proposed by WWF (Olson and Dinerstein 1994) that separately assessed the conservation potential and threat status of ecoregions. Having subsequently benefited from numerous insightful critiques, we feel the current approach now contains significant advances over the earlier method. Incorporating conservation potential and threat into the single conservation status assessment is the most prominent change.

The transparent and flexible design of this method should facilitate future efforts to reexamine results and analyze relationships among variables. By selectively removing or aggregating criteria, it is possible to test for correlation among variables, identify relationships among factors that are unique to particular METs or MHTs, and determine strong predictors for conservation status assessments.

GIS-Based Analyses

The parameters used in the conservation status assessment can be accurately measured using GIS technology when adequate digital data bases are available. Although many digital databases of landscape parameters were consulted in the conservation status assessments (e.g., remaining habitat data based on MSS and AVHRR satellite imagery), final values for the conservation status criteria were arrived at through assessments by regional experts.

In the vast majority of cases, the heightened accuracy available from a strictly GIS-based landscape analysis would be unlikely to change the conservation status assigned to ecoregions in this study because the categories used for each criterion are sufficiently broad. We also decided that there was no net benefit from applying rigid habitat classification standards to digital databases of uncertain quality and incomplete coverage. Currently, the available digital data on remaining natural habitats varies widely in terms of quality and availability throughout the LAC region. Moreover, our ability to effectively interpret habitat information based on remote sensing is currently poor for several major habitat types (e.g., deserts, grasslands).

We urge future researchers to apply GIS-based analysis of remaining habitats to test, update, and refine the results of this study, particularly as new and better-quality data become available. However, we are confident that the snapshot conservation status assessments made here have accurately placed the vast majority of ecoregions in the appropriate category.

Explanation of Criteria Used in This Analysis

Because the loss of biodiversity and alteration of ecological processes, both current and projected, are difficult to measure directly, conservation biologists are increasingly relying on landscape-level parameters as indicators. In this study, we employ the percent of original habitat lost, the presence of large blocks of intact original habitat, the degree of habitat fragmentation and degradation, the rate of conversion, and the degree of protection. Based on principles of landscape ecology and conservation biology (for detailed discussions see Noss 1992, Primack 1993, Meffe and Carroll 1994, Noss and Cooperrider 1994, Kareiva and Wennergren 1995), we assume that these variables help predict (a) an ecosystem's ability to maintain ecological processes (e.g., population and predator-prey dynamics within natural ranges of variation, pollination and seed dispersal, nutrient cycling, migration, dispersal, and gene flow); and (b) components of biodiversity (e.g., top predators or other keystone species) that will strongly influence how much and what kind of biodiversity will persist over the long term.

Although the first three criteria—habitat loss, habitat blocks, and fragmentation—can be highly correlated in some landscapes (e.g., continuing habitat fragmentation can reduce both the total area of habitat available and the size of habitat blocks), each of these criteria is analyzed separately because their relationships can be quite variable and different combinations can have substantially different ecological effects. For example, an ecoregion with 50 percent of its original habitat in two large blocks would have a higher probability of persistence than a similar habitat area that is highly fragmented into multiple small patches.

Total Habitat Loss

Habitat loss has been widely recognized as one of the primary factors contributing to the reduction and loss of terrestrial populations, species, and ecosystems. This criterion underscores the rapid loss of species predicted to occur in ecosystems when the total area of remaining habitat falls below minimum critical levels. Although there is no consensus on the mechanisms or exact thresholds for species loss in different ecosystems, both theoretical and empirical studies support the general correlation between habitat loss and species loss. Loss of habitat reduces biodiversity (a) by eliminating species or communities limited to particular geographic localities; (b) by decreasing the area of available original habitat below the minimum size needed to maintain critical, large-scale ecosystem dynamics; and (c) through the degradation and fragmentation of remaining habitats such that they become too small or isolated to individually or collectively support viable populations or maintain important ecological processes.

The second and third effects of habitat loss are reflected in large part in the criteria discussed below of habitat blocks and habitat fragmentation. Total habitat loss, measured at an ecoregional scale, reflects all these consequences but underscores the first and second. The loss of species due to elimination or truncation of habitats within their ranges is particularly important in ecoregions with high beta or gamma diversity.

Habitat Blocks

A critical parameter for assessing conservation status is the number and extent of blocks of contiguous habitat large enough for populations and ecosystem dynamics, each with different minimum area requirements, to function naturally. Large blocks of habitat sustain larger and more viable species populations, and they permit a broader range of species and ecosystem dynamics to persist. The geographic coverage of multiple large blocks also conserves a wider range of habitats, environmental gradients, and species.

The number of large blocks of habitat in different size categories is an important component of this criterion. Redundancy theory suggests that the presence of three or more examples of an ecosystem significantly increases its probability of long-term persistence. Factors such as fire, disease, pollution, deforestation or degradation can eliminate species or natural habitats within blocks. The presence of several blocks with similar communities allows recolonization and persistence of particular habitat types and species. Multiple habitat blocks that are well-distributed across the landscape are particularly important for conserving species and habitats in ecoregions that are characterized by a high degree of beta diversity (species turnover along environmental gradients).

The threshold size for viable blocks of habitat is broadly tailored to the scale of important ecosystem dynamics for different METs. In order to avoid misleading conclusions by applying continental size thresholds to island ecoregions (or very small continental or naturally disjunct systems), different sets of size thresholds are employed for each ecoregion size category. Scale problems are similarly addressed in the protected area criterion (see below).

Habitat Fragmentation

Persistent small population sizes are widely perceived as a major threat to conservation of terrestrial species. Habitat fragmentation places many low-density species in demographic jeopardy (Berger 1990; Laurance 1991; Newmark 1991; Wilcove et al. 1986). Fragmented ecosystems are stressed over a relatively large percentage of their intact habitat area by hunting pressure, fires from surrounding human activity, changes in microclimates, and invasion of exotic species (Lovejoy 1980; Saunders et al. 1991; Skole and Tucker 1993). As fragmentation increases, the amount of critical core habitat area decreases. Fragments under 100 km² are inadequate for maintaining viable populations of most large vertebrates. Some species of birds, trees, and butterflies that typically occur in very low densities, or have extremely patchy distributions, may also be lost in small fragments.[1] Ecoregions that still maintain large blocks (e.g., >1000 km²) of intact original habitat will at least have some core areas where large-scale ecological processes function naturally. The point values associated with different categories of this criterion reflect the greater severity of ecosystem disruption in landscapes where habitat fragmentation is more advanced (see Groom and Shumaker 1993).

Habitat Degradation

Habitat degradation resulting from human activities such as selective logging, pesticide exposure, burning, and overgrazing can have profound impacts on the long-term viability of ecosystems. However, degraded systems can be difficult or impossible to distinguish from pristine areas in remote sensing imagery. Moreover, quantifying habitat degradation at this scale is problematic because (a) habitat degradation is often patchy; (b) degraded states form a continuum and are not easily classified; and (c) the ecological effects of different forms of degradation are unclear and may occur on the scale of weeks to centuries. For these reasons, habitat degradation was not used as a criterion for most MHTs, although we suggest it be incorporated into future analyses when better data and interpretative methods become available.

Regional experts felt that habitat degradation would be particularly important in contributing to the conservation status of the grassland/savanna/shrubland and xeric formation METs, but were only comfortable in evaluating degradation for two MHTs: grasslands, savannas, and shrublands, and flooded grasslands. Therefore, for these two MHTs, the index value attributed to fragmentation was split between the fragmentation index and a degradation index.

Habitat Conversion

Conversion rates are less powerful estimators of conservation status than large-scale landscape features because (a) the actual ecological effects associated with conversion rates varies considerably depending on the original size of the ecoregion, the amount of remaining habitat when the rates were estimated, and spatial patterns of conversion; (b) the major uncertainties associated with current conversion rate estimates (see Whitmore and Sayer 1992); (c) the sensitivity of conversion rates to relatively minor changes in human behavior; and (d) the actual loss of habitat associated with recent estimates of habitat conversion, even for high rates, is typically small relative to the extensive landscape alteration of past centuries that is best reflected in the first three criteria. However, recent conversion rates do provide some information on the short-term trajectory of future habitat loss and fragmentation and are used to improve the accuracy of conservation status assessments rather than for estimating long-term threats. Our final conservation status analysis projects trends for habitat loss, fragmentation, and patch size into the next two decades and accounts for predicted or proposed large-scale conversion events (e.g., pending agricultural expansion projects or roads). After priority ecoregions have been identified, habitat conversion rates could be used to indicate which ecoregions would benefit most from certain kinds of conservation investments.

---

1. However, we also recognize that small fragments can be valuable for conserving representative communities and species (Shafer 1995), particularly in regions that are characterized by high levels of beta diversity. Many invertebrates, plants, fungi, and small vertebrates can be effectively conserved within small blocks of original habitat. Small fragments can also act as important stepping-stones for movement and dispersal of species.

Degree of Protection

The degree of protection criterion assesses how well humans have conserved sufficiently large blocks of intact habitat. Protected areas managed primarily for the conservation of biodiversity, or which otherwise effectively protect intact habitat, are emphasized in this criterion. Protected areas are not used as primary indicators for the conservation status of ecoregions because (a) the distribution of protected areas do not necessarily reflect the current extent and configuration of original habitat or the integrity of ecosystems over the entire landscape; (b) many protected areas encompass habitats that would not be considered intact; and (c) most protected areas are currently too few and small to encompass complete ecosystems and will only be effective if the surrounding landscape is well managed for biodiversity conservation.

One could potentially emphasize a lack of formally protected areas in the threat analysis (leading to the final conservation status) rather than considering their presence as a predictor of the snapshot conservation status. However, a lack of protected areas may not threaten some ecoregions due to their inaccessibility or harsh environments. Assessing threats using negative criteria (i.e., absence of protected areas) increases the probability of making a poor decision compared to basing conclusions on existing parameters.

Several important aspects should be considered in a comprehensive analysis of protected areas:

- The degree to which large remaining blocks of habitat are adequately protected within a system of protected areas
- The level of redundancy of protected areas that is needed to help ensure the long-term persistence of habitat types, communities, endangered species, or critical habitats for species or ecological processes
- The degree to which representative habitat types, communities, ecological gradients, endangered species, or critical habitats for resident or migratory species or ecological processes are contained within a system of protected areas
- The degree of connectivity among reserves for the dispersal of species and contiguity of large-scale ecosystem processes
- The effectiveness of management of protected areas and the ability of managers to defend protected areas based on their landscape configurations.

The first two considerations are addressed in the degree of protection criterion used here, while the latter three are best considered in more detailed analyses at the intra-ecoregion level.

Determination of Snapshot Conservation Status: Weighting and Categories

The parameters described in more detail in the following section were used to assess a snapshot conservation status index for each ecoregion. A threat analysis subsequently modifies the results, if necessary, to produce the final conservation status for each ecoregion.

The conservation status index has a point range from 0 to 100, with higher values denoting a higher level of endangerment. The range from 0 to 100 was deemed appropriate because previous experience with prioritizing and ranking large numbers of conservation areas has shown that a narrower index scale (e.g., from 0 to 30 as tested in the WWF Russia Biodiversity Project) yields little discrimination among units.

We feel that several landscape parameters should be given greater weight in the determination of the snapshot conservation status index: total habitat loss, habitat blocks, and degree of habitat fragmentation. The weighting of the different parameters in the index is:

Weight (percent)	Parameter
40	Total habitat loss
20	Habitat blocks
20	Habitat fragmentation (or, for grassland, savanna, and shrubland, and flooded grassland MHTs: 10 percent fragmentation and 10 percent habitat degradation)
10	Habitat conversion
10	Degree of protection

The point thresholds for different categories of conservation status are listed below (classification as "Extinct" is based on expert assessment):

Points	Conservation Status
0–6	Relatively Intact
7–36	Relatively Stable
37–64	Vulnerable
65–88	Endangered
89–100	Critical

Determination of Point Values for Each Criterion

Within each criterion, the determination of point values is intended to reflect real biological processes or the relative contribution of a particular situation to long-term biodiversity conservation. For example, both empirical evidence and theoretical ecology suggest that species loss and secondary extinctions increase dramatically with extensive habitat loss and fragmentation (Simberloff 1992; Terborgh 1992). In some cases, the assigned point values closely reflect the relationship between the causal factor and the ecological response (i.e., plotting of the points will approximate the curve of the general relationship). More subjective criteria are divided into broad categories that facilitate classification.

Certainly, alternative classification systems are possible. The original databases are provided in this report for those who may wish to reanalyze the data using different indices, criteria, weightings, or index ranges.

For most of the criteria, some point values other than those defined for the various categories were allowed in instances when regional experts deemed them useful and unavoidable.

Total Habitat Loss

This and the following criteria require a definition of what constitutes intact habitat. Although it would be preferable to classify habitats into several categories reflecting different levels of habitat alteration and degradation, constraints imposed by data availability and time and resource limitation necessitated using two broad classes—*intact* and *altered*.

We propose that intact, or remaining, habitat represents relatively undisturbed areas that are characterized by the maintenance of most original ecological processes and by communities with most of their original suite of native species. The following are some further clarifications of our definition for two broad types of habitat:

- Tropical broadleaf forest and conifer/temperate broadleaf forest METs: Canopy disturbance through human activities such as logging or small-scale agriculture is restricted to less than 10 percent of defined habitat block. Understory largely undisturbed by timber extraction, grazing, agriculture, or human caused fires. Although large mammals and birds may presently be absent from some blocks of habitat due to exploitation or insufficient area, such blocks may still sustain many native plant, invertebrate, and vertebrate species and their associated ecological processes.

- Grassland/savanna/shrubland and xeric formation METs: habitat has not been plowed or affected by major changes in flooding or surface water patterns. The vast majority of native plant species are still present in abundances within their natural range of variation and succession patterns follow natural cycles (e.g., grazing by domestic livestock and human-caused fires have no significant impact on native biota). Although large mammals and birds may presently be absent from some blocks of habitat due to exploitation or insufficient area, such blocks may still sustain many native plant, invertebrate, and vertebrate species and their associated ecological processes.

Points	Original Habitat Lost
0	0–10 percent
10	10–24 percent
20	24–49 percent
32	50–89 percent
40	> 90 percent

Habitat Blocks

In addition to using a different set of criteria for different MHTs, variables for this and the protected area criterion differ by ecoregion size category. In the tables for these two criteria, the smallest ecoregion size category (<100 km²) is also used for ecoregions, such as those on some Caribbean islands, that originally occurred in naturally disjunct patches of this size.

In the following tables for habitat block analysis (Tables A-1 to A-4), the information in the cells, unless stated otherwise, refers to the minimal requirement for at least one block of intact habitat. For an ecoregion within any given column, corresponding to its size, the table should be read from top to bottom until a statement true of the ecoregion is reached. The text ">500" should be interpreted as: "the ecoregion contains at least one block of intact habitat greater than 500 km²." Percentage values refer to the portion of the original ecoregion size that is still considered to be intact habitat.

At the BSP workshop, some regional experts felt that the minimum size and number of habitat blocks should be increased for very large ecoregions to better address the greater proportional amount of habitat necessary to conserve the range of biodiversity in extensive ecoregions. We acknowledge that some of the values above may represent minimally adequate conditions for some conservation status categories.

Table A-1. Habitat Block Analysis for Tropical Broadleaf Forest MET

	Ecoregion Size			
Point Value	>3,000 km²	1,000–3,000 km²	100–1,000 km²	<100 km²
2	>3,000 or ≥3 blocks >1000	>1,000 or ≥3 blocks >500	>500	80–100 percent intact
5	>1,000	>500	>250	40–80 percent intact
10	>500	≥3 blocks >250	≥3 blocks >100	10–40 percent intact
15	>250	>250	>100	1–10 percent intact
20	None >250	None >250	None >100	<1 percent intact

Note: The value ">1,000" means "the ecoregion contains at least one habitat block greater than 1,000 km²." For an ecoregion of any given size, this and the following tables should be read from top to bottom until a statement is reached that is true of the ecoregion.

Table A-2. Habitat Block Analysis for Conifer/Temperate Broadleaf Forest MET

	Ecoregion Size			
Point Value	>3,000 km²	1,000–3,000 km²	100–1,000 km²	<100 km²
2	>2,000 or ≥3 blocks >800	>800 or ≥3 blocks >500	>500	80–100 percent intact
5	>800	>500	>250	40–80 percent intact
10	≥3 blocks >250	≥3 blocks >250	≥3 blocks >100	10–40 percent intact
15	>250	>250	>100	1–10 percent intact
20	None >250	None >250	None >100	<1 percent intact

Table A-3. Habitat Block Analysis for Grassland/Savanna/Shrubland MET

	Ecoregion Size		
Point Value	>3,000 km²	1,000–3,000 km²	<1,000 km²
2	>750 or ≥3 blocks >500	>500 or ≥3 blocks >250	80–100 percent intact
5	>500	>250	40–80 percent intact
10	>250	≥2 blocks >100	10–40 percent intact
15	>100	>100	1–10 percent intact
20	None >100	None >100	<1 percent intact

Table A-4. Habitat Block Analysis for Xeric Formation MET

	Ecoregion Size		
Point Value	>3,000 km²	1,000–3,000 km²	<1,000 km²
2	≥2 blocks >500 or ≥3 blocks >200	>500 or ≥2 blocks >200	80–100 percent intact
5	>500	>200	40–80 percent intact
10	>200	≥2 blocks >100	10–40 percent intact
15	>100	>100	1–10 percent intact
20	None >100	None >100	<1 percent intact

Habitat Fragmentation

WWF has developed several fragmentation indices (Olson et al. 1994) that can be used to assess properties such as (a) the proportion of core habitat (habitat unaffected by edge effects using a predetermined effect distance) in a fragmented landscape; (b) the relative isolation of local fragments (multi-directional) based on inter-fragment distances relevant for most species and ecological processes; (c) the relative isolation of fragment clusters based on long distance dispersal abilities of vagile species; and (d) the influence of land use in areas between the fragments and type of habitat on the ecological isolation of fragments or fragment clusters. These GIS-based fragmentation analyses were not used in this study for the reasons discussed earlier in this appendix. Instead, we relied on regional experts' classification of ecoregions into broad fragmentation scenarios (humans are actually quite good at assessing fragmentation scenarios relative to computers).

Points	Degree of Habitat Fragmentation
0	*Relatively contiguous*: high connectivity; fragmentation low; long-distance dispersal along elevational and climatic gradients still possible.
5	*Low*: higher connectivity; more than half of all fragments clustered to some degree (i.e., have some degree of interaction with other intact habitat blocks).
12	*Medium*: intermediate connectivity; fragments somewhat clustered; intervening landscape allows for dispersal of many taxa through some parts of ecoregion.
16	*Advanced*: low connectivity; more larger fragments than in High category; fragments highly isolated; intervening landscape precludes dispersal for most
20	*High*: most fragments small and or noncircular; little core habitat due to edge effects (e.g., extending for 0.75–1.0 km for physical edge effects and for 40 km for hunting pressure); most individual fragments and clusters of fragments highly isolated; intervening landscapes preclude dispersal for most taxa.

In ecoregions that are comprised of several naturally disjunct areas, fragmentation should be assessed within individual remaining blocks (the largest blocks contribute the most to the assessment) and not on the basis of connectivity between blocks.

Habitat Degradation

Grassland, savanna, and shrubland, and flooded grassland MHTs were assessed for both fragmentation and degradation. For these MHTs, fragmentation receives half the points allotted above for each category, and habitat degradation is rated as follows:

Points	Degree of Habitat Degradation
0	*Low*: populations of native plant species, successional processes, and disturbance regimes relatively unaffected by anthropogenic grazing and burning.
5	*Medium*: population of native plant species persist in reduced numbers; succession and disturbance processes modified.
10	*High*: few native plant species persist; large native herbivores eliminated; succession and disturbance processes significantly altered.

Habitat Conversion

For each ecoregion, we estimated the recent rate of habitat conversion (i.e., proportion of the remaining habitat in an ecoregion being converted from intact to altered habitat per year). We tried to estimate rates relevant for the previous five-year period.

Points	Conversion per Annum
0	<0.5 percent
6	0.5–2.0 percent
8	2.1–3 percent
9	3.1–4 percent
10	>4 percent

Future large-scale conversions predicted on the basis of impending development projects, exploitation plans, colonization, or other factors are addressed in the threat analyses.

Degree of Protection

As for the habitat block criterion, Tables A-5 to A-7 should be read from top to bottom, within any given size category, until a statement is reached that is true for the ecoregion. A definition such as ">500" should be interpreted as "the ecoregion contains at least one protected area which includes a block of intact habitat greater than 500 km². " Percentage values refer to the portion of remaining intact habitat incorporated into a protected area system.

At the BSP workshop, some regional experts felt that only protected areas with a staff of qualified managers, transportation, and a budget should be used in this analysis. Some experts also felt that a different set of criteria should be used for very large ecoregions. Future analyses should probably incorporate such a refinement; Table A-7 suggests a possible set of criteria that were suggested by regional experts but which, however, were not used to obtain the final values in this study.

Table A-5. Degree of Protection Analysis for Broadleaf and Conifer Forest METs

	Ecoregion Size			
Point Value	>3,000 km²	1,000–3,000 km²	100–1,000 km²	<100 km²
1	≥2 areas >1,000	≥2 areas >750	≥3 areas >250	>50 percent protected
4	>500	>500	>250	40–50 percent protected
6	>250	>250	>100	20–40 percent protected
8	>100	>100	Areas exist but none >100	1–20 percent protected
10	None >100	None >100	No areas	No areas

Note: The value ">1,000" means "ecoregion contains at least one protected area containing a block of intact habitat greater than 1,000 km²." For an ecoregion of any given size, this and the following tables should be read from top to bottom until a statement is reached that is true of the ecoregion.

Table A-6. Degree of Protection Analysis for Grassland/Savanna/Shrubland and Xeric Formation METs

	Ecoregion Size			
Point Value	>3,000 km²	1,000–3,000 km²	100–1,000 km²	<100 km²
1	≥2 areas >750	≥2 areas >500	≥2 areas >250	>50 percent protected
4	>500	>500	>250	40–50 percent protected
6	>250	>250	>100	20–40 percent protected
8	>100	>100	Areas exist but none >100	1–20 percent protected
10	None >100	None >100	No areas	No areas

Table A-7. Degree of Protection Analysis Suggested for Large Ecoregions

	Ecoregion Size		
Point Value	>500,000 km²	200,000-500,000 km²	50,000–200,000 km²
1	>8,000	>4,000	>2,000
4	≥3 areas >4,000	≥3 areas >2,000	≥3 areas >1,000
6	>2,000	>1,000	>500
8	>1,000	>500	>250
10	None >1,000	None >500	None >250

Assessing Final Conservation Status

The snapshot conservation status of ecoregions was revised on the basis of a threat analysis to develop final conservation status assessments. For example, some forests may be largely intact and would receive a snapshot status of Relatively Stable, but the imminence of extensive logging concessions might warrant shifting the ecoregion's final conservation status to Vulnerable. The final conservation status assessments should reflect the urgency of conservation action as well as the ecological integrity of ecoregions.

To evaluate as objectively as possible the threats facing an ecoregion over about the next 20 years, we estimated a threat value for each ecoregion of high, medium or low. Ecoregions with high threat estimates were moved up one class in their conservation status ranking (e.g., from Vulnerable to Endangered). Medium threat estimates influenced the rankings depending upon our judgment. Low threat estimates were not used to modify snapshot conservation status rankings.

At the BSP Workshop, experts qualitatively determined high, medium and low evaluations. A more quantitative method, using points, was generally used at the subsequent WWF LAC Program Workshop. This point system is outlined below. For the most part, the latter determinations are those that are included in this report.

Method of Threat Analysis

The major threats to each ecoregion in terms of their type, intensity, and timeframe are identified here. Threat analyses are inherently complex since factors may effect ecosystems directly or indirectly, and there are numerous and not well understood synergistic interactions among factors. Thus, the formal analysis of threats used here is based on the overall effect of threats on habitat modification and ecosystem degradation, regardless of the number or type of threats identified.

We used an index of 0-100 points to determine pending threats to the ecoregion. Points were attributed to three major types of threat: conversion threats (maximum of 50 points), degradation threats (maximum of 30), and wildlife exploitation threats (maximum of 20 points). High threat was considered equivalent to 70-100 total points, medium threat equivalent to 20-69, and low threat to 0-19 points.

The following table describes how points were allocated within each threat category.

Points	Degree of Threat
Conversion Threats	
0	No conversion threats recognized for ecoregion.
10	Threats may significantly alter between 5 percent and 9 percent of remaining habitat within 20 years.
20	Threats may significantly alter between 10 percent and 24 percent of remaining habitat within 20 years.
50	Threats may significantly alter 25 percent or more of remaining habitat within 20 years.
Degradation Threats	
0	No degradation threats recognized for ecoregion.
15	Populations of native plant species experiencing significant mortality and poor recruitment due to degradation factors. Succession and disturbance processes modified. Some abandonment and underuse of seasonal, migratory, and breeding movements by species. Pollutants and linked effects commonly found in target species or assemblages.
30	Many populations of native plant species experiencing high mortality and low recruitment due to degradation factors. Succession and disturbance processes significantly altered. Low habitat quality for sensitive species. Abandonment and disruption of seasonal, migratory, or breeding movements. Pollutants and linked effects widespread in ecosystem (e.g., recorded in several trophic levels).
Wildlife Exploitation	
0	No wildlife exploitation recognized for ecoregion.
10	Moderate levels of wildlife exploitation; populations of game or trade species persisting but in reduced numbers.
20	High intensity of wildlife exploitation in region with elimination of local populations of most target species imminent or complete.

The bulleted list that follows describes the type of threats in each category.

Conversion threats
- intensive logging and associated road building
- intensive grazing
- agricultural expansion, plantations, and clearing for development
- permanent alteration from burning

Degradation threats
- pollution (e.g., oil, pesticides, herbicides, mercury, heavy metals, defoliants)
- burning
- introduced species

- firewood extraction
- grazing
- unsustainable extraction of non-timber products
- road building and associated erosion and land-slide damage
- off-road vehicle damage
- selective logging
- excessive recreational impacts

Wildlife exploitation
- hunting and poaching
- unsustainable extraction of wildlife and plants as commercial products
- harassment and displacement by commercial and recreational users

Appendix B

Methods Used for Assessing the Conservation Status of Mangrove Units

The assessment of mangroves took place at the LAC Mangroves Workshop held by WWF November 2-4, 1994. The method presented here is much abbreviated given that the general approach is discussed in detail for terrestrial ecoregions. Unless stated otherwise, refer to Appendix A for details on the justification and application of the method. A full report on the results of the mangrove workshop will be eventually published by WWF.

Determination of Snapshot Conservation Status: Weighting and Categories

Two landscape parameters were given greatest weight in the conservation status index: loss of original habitat and habitat blocks. The relative contribution of parameters used in the index is:

Weight (percent)	Parameter
40	Total habitat loss
20	Habitat blocks
10	Degree of protection
10	Habitat conversion
10	Water quality and hydrographic integrity
5	Habitat fragmentation
5	Degree of alteration of catchment basin

The point thresholds for different categories of conservation status are listed below (classification as "Extinct" is based on expert assessment):

Points	Conservation Status
0–6	Relatively Intact
7–36	Relatively Stable
37–64	Vulnerable
65–88	Endangered
89–100	Critical

Determination of Point Values for Each Criterion

Total Habitat Loss

The following definition of intact mangrove habitat is used. Canopy opening through logging or other human disturbance is restricted to less than 10 percent of defined habitat block. No major changes in hydrology or severe pollution immediately threaten persistence of habitat block. Understory largely undisturbed from draining, timber extraction, grazing, agriculture, bark harvesting, or human-caused fires. Although large mammals, birds, and reptiles may presently be absent from some blocks of habitat due to exploitation or insufficient area, such blocks may still sustain many native plant, invertebrate, and vertebrate species and their associated ecological processes.

Points	Original Habitat Lost
0	0–10 percent
10	10–24 percent
20	24–49 percent
32	50–89 percent
40	> 90 percent

Habitat Blocks

Table B-1 presents the method used to determine point values for the habitat blocks criterion.

Table B–1. Habitat Block Analysis for Mangrove Units

Point Value	Unit Size		
	>3,000 km²	*1,000–3,000 km²*	*<1,000 km²*
2	>1000 or ≥3 blocks >500	>750 or ≥3 blocks >500	90–100 percent
5	>500	>500	70–90 percent
10	>250	≥3 blocks >200	40–70 percent
15	>100	>75	10–40 percent
20	None >100	None >75	<10 blocks

Note: The value ">500" means "the unit contains at least one habitat block greater than 500 km²." The value "90 percent" means "the unit contains at least one habitat block that is 90 percent the size of the largest original unit." For a unit of any given size, the table should be read from top to bottom until a statement is reached that is true of the unit.

Degree of Protection

The degree of protection analysis used for mangroves is presented in Table B-2.

Table B–2. Degree of Protection Analysis for Mangrove Units

Point Value	Unit Size		
	>1,000 km²	*250–1,000 km²*	*<250 km²*
1	≥3 areas >200	≥3 areas >100	>50 percent protected
4	>200	≥2 areas >100	40–50 percent protected
6	≥2 areas >100	>100	20–40 percent protected
8	>100	Areas exist but none >100	<20 percent protected
10	None >100	No areas	No areas

Note: The value ">200" means "the unit contains at least one protected area containing a block of intact habitat greater than 200 km²." For a unit of any given size, this table should be read from top to bottom until a statement is reached that is true of the unit.

Habitat Conversion

For each ecoregion, the recent rate of direct habitat conversion was estimated (i.e., proportion of the remaining original habitat within an ecoregion being converted from intact to altered habitat per year). We tried to estimate rates that are relevant for about the last five years.

Points	Conversion per Annum
0	<0.5 percent
6	0.5–2.0 percent
8	2.1–3.0 percent
9	3.1–4.0 percent
10	>4.0 percent

Future large-scale conversions predicted on the basis of impending development projects, exploitation plans, colonization, or other factors were addressed in the threat analyses.

Water Quality and Integrity of Hydrographic Processes

Mangrove ecosystems are known to be particularly sensitive to changes in water quality and hydrographic processes. The purpose of this criterion is to provide a means to reassess the amount of available high quality habitat estimated from the habitat loss and large habitat blocks assessments. Even standing mangrove forests may soon represent inviable habitat due to significant changes in water quality or hydrographic processes. Important parameters are pH, turbidity, dissolved oxygen, pesticides, heavy metals, suspended solids, hydrocarbons (oil), and alteration of tidal cycles and freshwater discharge. The following categories were used:

Points	Remaining Habitat Affected by Changes
2	0–20 percent
4	21–40 percent
6	41–60 percent
8	61–80 percent
10	81–100 percent (i.e., assumed loss of mangrove without significant restoration)

Habitat Fragmentation

Mangrove ecoregions are comprised of naturally disjunct habitats occurring along the coastlines of continents and islands. Many habitat units themselves are highly fragmented due to channels and the patchy and dynamic nature of suitable substrate. Mangroves also disperse well and are able to colonize new areas relatively rapidly. For these reasons, mangrove forests and their associated biota are, with some exceptions, not particularly sensitive to habitat fragmentation as long as natural hydrographic conditions persist.

Points	Degree of Fragmentation
0	*Relatively contiguous*: high connectivity; fragmentation low; long-distance dispersal of mangrove specialists still possible.
1	*Low*: higher *connectivity*; more than half of all fragments clustered to some degree (i.e., have some degree of interaction with other intact habitat blocks).
3	*Medium*: intermediate connectivity; fragments *somewhat* clustered; intervening landscape allows for dispersal of many taxa through some parts of ecoregion.
4	*Advanced*: low *connectivity*; more larger fragments than in High category; fragments highly isolated; intervening land scape inhibits dispersal for most taxa.
5	*High*: most fragments small and or noncircular; little core habitat due to edge effects (e.g., extending for 0.75-1.0 km for physical edge effects and for 40 km for hunting pressure); most individual fragments and clusters of fragments highly isolated; intervening landscapes preclude dispersal for most taxa.

Fragmentation was assessed within individual blocks and not on the basis of connectivity between blocks. The state of the largest blocks contributes relatively more to the assessment.

Degree of Alteration of the Catchment Basin

Changes in catchment basins that can affect mangroves include alteration of the freshwater inflow from rivers and runoff, increased sedimentation, increased pollution, increased access to hunters and loggers, and loss of surrounding wildlife populations and resource areas for mangrove associated species.

Points	Catchment Basin Altered
0	0–19 percent
2	20–39 percent
3	40–59 percent
4	60–89 percent
5	81–100 percent

Assessing Final Conservation Status

The point totals for each of the landscape parameters were summed and a snapshot conservation status assigned to each mangrove unit (as in Appendix A). The snapshot conservation status assessments were subsequently modified by a threat analyses to better reflect the long-term trajectory of the mangrove unit based on significant, large-scale threats.

Method of Threat Analysis

Suggested categories for threats to mangroves include:

Conversion threats
- logging and charcoal production
- agricultural expansion and clearing for development
- salt extraction
- shrimp farming
- bark harvesting
- draining or channelization (alteration of tidal or freshwater regimes)

Degradation threats
- pollution (e.g., oil, pesticides, heavy metals, defoliants)
- burning
- introduced species
- firewood extraction

Wildlife exploitation
- hunting, wildlife trade, and overfishing

As in Appendix A, we used an index of 0-100 points to determine pending threats to the unit. High threat was considered equivalent to 70-100 points, medium threat equivalent to 20-69, and low threat to 0-19 points.

Modification of the snapshot conservation status of mangrove units followed the same decision rules used for terrestrial ecoregions. The table below describes how points were allocated within each threat category.

Points	Description
Conversion Threats	
0	No conversion threats recognized for ecoregion.
10	Threats may significantly alter between 5 percent and 9 percent of remaining habitat within 20 years.
20	Threats may significantly alter between 10 percent and 24 percent of remaining habitat within 20 years.
50	Threats may significantly alter 25 percent or more of remaining habitat within 20 years.
Degradation Threats	
0	No degradation threats recognized for ecoregion.
15	Populations of native plant species experiencing significant mortality and poor recruitment due to degradation factors. Succession and disturbance processes modified. Pollutants and linked effects commonly found in target species or assemblages.

Points	Description
Degradation Threats (continued)	
30	Many populations of native plant species experiencing high mortality and low recruitment due to degradation factors. Succession and disturbance processes significantly altered. Pollutants and linked effects widespread in ecosystem (e.g., recorded in several trophic levels).
Wildlife Exploitation	
0	No wildlife exploitation recognized for ecoregion.
10	Moderate levels of wildlife exploitation, populations of game or trade species persisting but in reduced numbers
20	High intensity of wildlife exploitation in region with elimination of local populations of most target species imminent or complete.

Appendix C

Definitions of Major Ecosystem Types and Major Habitat Types

Tropical Broadleaf Forest MET

General Characteristics: These are closed canopy tropical forests dominated by broadleaf species. Four distinguishing characteristics are (a) drip tips on leaves; (b) smooth columnar boles of trees; (c) presence of ant-plants; and (d) presence of true lianas. Other characteristics of these forests are: high species richness, particularly in wet, aseasonal environments; propensity of many taxa toward restricted geographic distributions (endemism) and high habitat specificity; propensity toward a high degree of beta diversity, particularly in wetter, aseasonal environments and in more complex environments; seed dispersal and pollination largely by animals, sometimes over large distances; high numbers of obligate pollinators; high proportion of tightly linked ecological interactions (e.g., symbioses); many tree, vertebrate, and invertebrate species occurring at relatively low densities (i.e., large areas needed for viable populations); high area requirements for some species due to patchy resource distribution (e.g., frugivores); and importance of keystone predators and resource species (e.g., fig trees) for maintaining ecosystem integrity.

Sensitivity to Disturbance: Secondary extinctions and changes in species compositions are predictable as a result of loss of keystone predators and resource species. Many taxa, particularly understory species, are sensitive to changes in microclimate. Movement of many species is restricted to intact, closed canopy forest. Widespread forest clearance can significantly alter soil structure and successional patterns. Ecosystem processes and populations significantly affected by habitat fragmentation due to sensitivities discussed above.

Conservation Considerations: Tropical broadleaf forests typically require large protected areas to maintain viable populations, sustain ecological processes, and buffer core habitats from hunting and other edge-related disturbances. High levels of beta diversity may warrant numerous, well-dispersed reserves in many cases. Biota very sensitive to edge effects and isolation—linkages among habitat blocks should be of closed canopy forest. Altitudinal linkages important for seasonal movements of many bird, mammal, and invertebrate species.

Tropical Moist Broadleaf Forest MHT

The general criterion used to define this MHT was that less than 50 percent of canopy tree species be deciduous.

Tropical Dry Broadleaf Forest MHT

The general criterion used to define this MHT was that more than 50 percent of canopy tree species be deciduous.

Conifer/Temperate Broadleaf Forest MET

General Characteristics: Forests are composed of conifers, mixed conifer/broadleafs, or temperate broadleaf species. Characterized by lower levels of species richness, endemism, and beta diversity than tropical broadleaf forests, but regional and local range-restricted species common in some regions. Narrow altitudinal zonation of communities common. Wind pollination common. Tree species occur at higher individual densities. Periodic disturbances can often affect large areas in some forest types (e.g., fire, epizootics). Altitudinal linkages important for seasonally migrating species. Riparian and bottomland communities often are critical climatic refuges and resource habitats for wildlife. Species are generally better adapted for dispersal across non-forest habitats than tropical broadleaf species. Understory species sensitive to overgrazing and frequent burning.

Conservation Considerations: Large reserves are required to maintain viable populations of larger predators and to help ensure that some undisturbed habitat remains after large disturbance events to act as refuges for wildlife and source pools for species and to maintain ecological processes. Habitat linkages along altitudinal gradients are critical as is conservation of riparian and bottomland habitat. Grazing and anthropogenic fires should be strictly controlled.

Temperate Forest MHT

The forests of the southern cone of South America are the only representatives of this MHT in Latin America and the Caribbean.

Tropical and Subtropical Coniferous Forest MHT

Some taxa display pronounced endemism and narrow altitudinal zonation. Isolated montane forests may require numerous, well-dispersed reserves for conservation of their biodiversity.

Grassland/Savanna/Shrubland MET

General Characteristics: This category encompasses a wide range of community types with various combinations of forest, grassland, shrublands, and woodlands. Their shared features include: many species of plants and animals widely distributed within ecoregions; communities relatively resilient to short-term disturbances but not to prolonged burning or overgrazing; and sensitivity to fragmentation (i.e., from habitat loss and fencing), which disrupts movement patterns of larger vertebrates that rely on patchy or seasonal resources.

Conservation Considerations: Small reserves may effectively conserve plant and invertebrate species but need to be large enough to avoid complete alteration by fire or edge effects. Larger reserves needed to conserve intact vertebrate faunas, particularly for species tracking patchy or seasonal resources. Hunting, burning, and grazing need to be strictly

controlled. Riparian areas and sources of water are critical for many species. Introduced species can have significant ecological impacts.

Grassland, Savanna, and Shrubland MHT

Characteristics of this MHT are as given for the MET above.

Flooded Grassland MHT

These areas are regularly inundated by water and are particularly sensitive to pollution and changes in patterns of flooding from draining, channelization, or alteration of surrounding habitat.

Montane Grassland MHT

Some montane communities regenerate slowly and may have low resiliency to disturbances such as grazing or burning. Plant species may have patchy distributions due to altitudinal zonation or other specific habitat requirements. Larger vertebrates may require intact altitudinal corridors for seasonal movements or large reserves to maintain sufficient resources in patchy environments.

Xeric Formation MET

General Characteristics and Conservation Considerations: Xeric formations can be characterized by relatively rich floras and a propensity for endemism in some taxa. These communities are sensitive to overgrazing, burning, invasion by exotics, and clearing. Many species sensitive to fragmentation and the loss of critical habitats such as riparian zones and waterholes. Populations of larger vertebrates often occur at low densities.

Mediterranean Scrub MHT

These scrublands with Mediterranean-type climates are found only in two areas of LAC: central Chile and the coast of California. They are characterized by: high species richness, particularly for plants and invertebrates; very high levels of beta diversity and endemism (both local and regional); sensitivity to overgrazing and exotic species; and sensitivity of many vertebrate species to fragmentation. Small reserves may be effective for conserving many components of biodiversity, particularly plants and invertebrates, but they must be large enough to be resilient to disturbance (e.g., retain some undisturbed habitat after periodic fire events).

Desert and Xeric Shrubland MHT

Characteristics of this MHT are as given for the MET above.

Restinga MHT

Restingas refer to habitats of xeromorphic coastal dune or sandy soil vegetation, ranging from low herbaceous ground cover to open forests. They are extensive in some areas with rich floras and many endemics and are sensitive to overgrazing, burning, and disturbance of dunes.

Mangrove MET

General Characteristics and Conservation Considerations: The word mangrove is used to describe a salt-tolerant forest ecosystem that occupies sheltered tropical and subtropical coastal estuarine, lagoon, deltaic, or carbonate platform environments. Although the constituent plant species (also called mangroves) are not closely related, they have common morphological, physiological and reproductive adaptations that allow survival and development in very saline, waterlogged and reduced substrates. These substrates are often poorly consolidated and subject to rapid changes.

Mangrove forest ecosystems receive inputs of matter and energy from both land and sea. These inputs include fresh water, sediments and nutrients from land and tidal flushing and saline intrusions from the sea. These inputs act as energy subsidies that increase the performance of the system and help maintain high rates of organic matter fixation and active ecosystem processes.

Forests growing in areas periodically inundated by brackish or marine water and comprised of trees in the genera *Avicennia, Conocarpus, Laguncularia, Pelliciera,* and *Rhizophora*; high species richness when associated marine and freshwater species are considered; low endemism and beta diversity; habitat fragmentation not a major problem at local scales because of rapid regeneration of mangroves (as long as necessary hydrographic conditions are still present) and adaptation of biota to patchiness of habitat; mangroves are particularly sensitive to changes in hydrography (e.g., draining or changes in tidal or river outflow patterns) and pollution; surrounding forests often contain important resources for terrestrial wildlife.

Appendix D

Hierarchical Classification Scheme of LAC Ecoregions

MAJOR ECOSYSTEM TYPE

Major Habitat Type

Bioregion
Ecoregion—countries in which ecoregion occurs (ecoregion numbers correspond to those on the large-format map)

TROPICAL BROADLEAF FORESTS

Tropical Moist Broadleaf Forests

Caribbean
1. Cuban moist forests—Cuba
2. Hispaniolan moist forests—Haiti, Dominican Republic
3. Jamaican moist forests—Jamaica
4. Puerto Rican moist forests—Puerto Rico
5. Windward Islands moist forests—Windward Islands
6. Leeward Islands moist forests—Leeward Islands

Central America
7. Oaxacan moist forests—Mexico
8. Tehuantepec moist forests—Mexico, Guatemala, Belize
9. Yucatán moist forests—Mexico
10. Sierra Madre moist forests—Mexico, Guatemala, El Salvador
11. Central American montane forests—Mexico, Guatemala, El Salvador, Honduras
12. Belizean swamp forests—Belize
13. Central American Atlantic moist forests—Guatemala, Belize, Honduras, Nicaragua, Costa Rica, Panama

14. Costa Rican seasonal moist forests—Costa Rica, Nicaragua
15. Isthmian–Pacific moist forests—Costa Rica, Panama
16. Talamancan montane forests—Costa Rica, Panama

Orinoco
17. Cordillera La Costa montane forests—Venezuela
18. Orinoco Delta swamp forests—Venezuela, Guyana
19. Trinidad & Tobago moist forests—Trinidad & Tobago
20. Guianan Highlands moist forests—Venezuela, Brazil, Guyana
21. Tepuis—Venezuela, Brazil, Guyana, Suriname, Colombia

Amazonia
22. Napo moist forests—Peru, Ecuador, Colombia
23. Macarena montane forests—Colombia
24. Japura/Negro moist forests—Colombia, Venezuela, Brazil, Peru
25. Uatama moist forests—Brazil, Venezuela, Guyana
26. Amapá moist forests—Brazil, Suriname
27. Guianan moist forests—Venezuela, Guyana, Suriname, Brazil, French Guiana
28. Paramaribo swamp forests—Suriname
29. Ucayali moist forests—Brazil, Peru
30. Western Amazonian swamp forests—Peru, Colombia
31. Southwestern Amazonian moist forests—Brazil, Peru, Bolivia
32. Juruá moist forests—Brazil

33. Varzea forests — Brazil, Peru, Colombia
34. Purus/Madeira moist forests — Brazil
35. Rondônia/Mato Grosso moist forests — Brazil, Bolivia
36. Beni swamp and gallery forests — Bolivia, Brazil
37. Tapajós/Xingu moist forests — Brazil
38. Tocantins moist forests — Brazil

Northern Andes
39. Chocó/Darién moist forests — Colombia, Panama
40. Eastern Panamanian montane forests — Panama, Colombia
41. Northwestern Andean montane forests — Colombia, Ecuador
42. Western Ecuador moist forests — Ecuador, Colombia
43. Cauca Valley montane forests — Colombia
44. Magdalena Valley montane forests — Colombia
45. Magdalena/Urabá moist forests — Colombia
46. Cordillera Oriental montane forests — Colombia, Venezuela
47. Eastern Cordillera Real montane forests — Ecuador, Colombia, Peru
48. Santa Marta montane forests — Colombia
49. Venezuelan Andes montane forests — Venezuela, Colombia
50. Catatumbo moist forests — Venezuela, Colombia

Central Andes
51. Peruvian Yungas — Peru
52. Bolivian Yungas — Bolivia, Argentina
53. Andean Yungas — Argentina, Bolivia

Eastern South America
54. Brazilian Coastal Atlantic forests — Brazil
55. Brazilian Interior Atlantic forests — Brazil, Argentina, Paraguay

Tropical Dry Broadleaf Forests

Caribbean
56. Cuban dry forests — Cuba
57. Hispaniolan dry forests — Haiti, Dominican Republic
58. Jamaican dry forests — Jamaica
59. Puerto Rican dry forests — Puerto Rico
60. Bahamian dry forests — Bahamas, Turks & Caicos Islands
61. Cayman Islands dry forests — Cayman Islands
62. Windward Islands dry forests — Windward Islands
63. Leeward Islands dry forests — Leeward Islands

Northern Mexico
64. Baja California dry forests — Mexico
65. Sinaloan dry forests — Mexico
66. Tamaulipas/Veracruz dry forests — Mexico

Central America
67. Jalisco dry forests — Mexico
68. Balsas dry forests — Mexico
69. Oaxacan dry forests — Mexico
70. Veracruz dry forests — Mexico
71. Yucatán dry forests — Mexico
72. Central American Pacific dry forests — El Salvador, Honduras, Nicaragua, Costa Rica, Guatemala
73. Panamanian dry forests — Panama

Orinoco
74. Llanos dry forests — Venezuela
75. Trinidad & Tobago dry forests — Trinidad & Tobago

Amazonia
76. Bolivian lowland dry forests — Bolivia, Brazil

Northern Andes
77. Cauca Valley dry forests — Colombia
78. Magdalena Valley dry forests — Colombia
79. Patía Valley dry forests — Colombia
80. Sinú Valley dry forests — Colombia
81. Ecuadorian dry forests — Ecuador
82. Tumbes/Piura dry forests — Ecuador, Peru
83. Marañón dry forests — Peru
84. Maracaibo dry forests — Venezuela
85. Lara/Falcón dry forests — Venezuela

Central Andes
86. Bolivian montane dry forests — Bolivia

CONIFER/TEMPERATE BROADLEAF FORESTS

Temperate Forests

Southern South America
87. Chilean winter-rain forests — Chile
88. Valdivian temperate forests — Chile, Argentina
89. Subpolar *Nothofagus* forests — Chile, Argentina

Tropical and Subtropical Coniferous Forests

Caribbean
- 90. Cuban pine forests—Cuba
- 91. Hispaniolan pine forests—Haiti, Dominican Republic
- 92. Bahamian pine forests—Bahamas, Turks & Caicos Islands

Northern Mexico
- 93. Sierra Juarez pine-oak forests—Mexico, U.S.
- 94. San Lucan pine-oak forests—Mexico
- 95. Sierra Madre Occidental pine-oak forests—Mexico, U.S.
- 96. Central Mexican pine-oak forests—Mexico
- 97. Sierra Madre Oriental pine-oak forests—Mexico
- 98. Veracruz pine-oak forests—Mexico

Central America
- 99. Mexican transvolcanic pine-oak forests—Mexico
- 100. Veracruz montane forests—Mexico
- 101. Sierra Madre del Sur pine-oak forests—Mexico
- 102. Central American pine-oak forests—Mexico, Guatemala, El Salvador, Honduras, Nicaragua
- 103. Belizean pine forests—Belize
- 104. Miskito pine forests—Nicaragua, Honduras

Eastern South America
- 105. Brazilian *Araucaria* forests—Brazil, Argentina

GRASSLANDS/SAVANNAS/SHRUBLANDS

Grasslands, Savannas, and Shrublands

Northern Mexico
- 106. Central Mexican grasslands—Mexico, U.S.
- 107. Eastern Mexican grasslands—Mexico

Central America
- 108. Tabasco/Veracruz savannas—Mexico
- 109. Tehuantepec savannas—Mexico

Orinoco
- 110. Llanos—Venezuela, Colombia

Amazonia
- 111. Guianan savannas—Suriname, Guyana, Brazil, Venezuela
- 112. Amazonian savannas—Brazil, Colombia, Venezuela
- 113. Beni savannas—Bolivia

Eastern South America
- 114. Cerrado—Brazil, Paraguay, Bolivia
- 115. Chaco savannas—Argentina, Paraguay, Bolivia, Brazil
- 116. Humid Chaco—Argentina, Paraguay, Bolivia, Brazil
- 117. Córdoba montane savannas—Argentina

Southern South America
- 118. Argentine Monte—Argentina
- 119. Argentine Espinal—Argentina
- 120. Pampas—Argentina
- 121. Uruguayan savannas—Uruguay, Brazil, Argentina

Flooded Grasslands

Caribbean
- 122. Cuban wetlands—Cuba
- 123. Enriquillo wetlands—Haiti, Dominican Republic

Northern Mexico
- 124. Central Mexican wetlands—Mexico

Central America
- 125. Jalisco palm savannas—Mexico
- 126. Veracruz palm savannas—Mexico
- 127. Quintana Roo wetlands—Mexico

Orinoco
- 128. Orinoco wetlands—Venezuela

Amazonia
- 129. Western Amazonian flooded grasslands—Peru, Bolivia
- 130. Eastern Amazonian flooded grasslands—Brazil
- 131. São Luis flooded grasslands—Brazil

Northern Andes
- 132. Guayaquil flooded grasslands—Ecuador

Eastern South America
- 133. Pantanal—Brazil, Bolivia, Paraguay
- 134. Paraná flooded savannas—Argentina

Montane Grasslands

Central America
- 135. Mexican alpine tundra—Mexico

136. Costa Rican paramo — Costa Rica

Northern Andes
137. Santa Marta paramo — Colombia
138. Cordillera de Mérida paramo — Venezuela
139. Northern Andean paramo — Colombia, Ecuador

Central Andes
140. Cordillera Central paramo — Ecuador, Peru
141. Central Andean puna — Bolivia, Argentina, Peru, Chile
142. Central Andean wet puna — Peru, Bolivia, Chile
143. Central Andean dry puna — Argentina, Bolivia, Chile

Southern South America
144. Southern Andean steppe — Argentina, Chile
145. Patagonian steppe — Argentina, Chile
146. Patagonian grasslands — Argentina, Chile

XERIC FORMATIONS

Mediterranean Scrub

Northern Mexico
147. California coastal sage-chaparral — Mexico, U.S.

Central Andes
148. Chilean matorral — Chile

Deserts and Xeric Shrublands

Caribbean
149. Cuban cactus scrub — Cuba
150. Cayman Islands xeric scrub — Cayman Islands
151. Windward Islands xeric scrub — Windward Islands
152. Leeward Islands xeric scrub — Leeward Islands

Northern Mexico
153. Baja California xeric scrub — Mexico
154. San Lucan mezquital — Mexico
155. Western Mexican mezquital — Mexico, U.S.
156. Sonoran xeric scrub — Mexico, U.S.
157. Northern Sonoran cactus scrub — Mexico, U.S.
158. Mexican Interior chaparral — Mexico, U.S.

159. Chihuahuan xeric scrub — Mexico, U.S.
160. Central Mexican mezquital — Mexico
161. Eastern Mexican matorral — Mexico
162. Eastern Mexican mezquital — Mexico, U.S.
163. Central Mexican cactus scrub — Mexico

Central America
164. Pueblan xeric scrub — Mexico
165. Guerreran cactus scrub — Mexico
166. Motagua Valley thornscrub — Guatemala

Orinoco
167. Aruba/Curaçao/Bonaire cactus scrub — Netherlands Antilles
168. La Costa xeric shrublands — Venezuela
169. Araya and Paría xeric scrub — Venezuela

Northern Andes
170. Galapagos Islands xeric scrub — Ecuador
171. Guajira/Barranquilla xeric scrub — Colombia, Venezuela
172. Paraguaná xeric scrub — Venezuela

Central Andes
173. Sechura desert — Peru, Chile
174. Atacama desert — Chile

Eastern South America
175. Caatinga — Brazil

Restingas

Northern Andes
176. Paraguaná restingas — Venezuela

Eastern South America
177. Northeastern Brazil restingas — Brazil
178. Brazilian Atlantic Coast restingas — Brazil

MANGROVES

Atlantic Mangrove Complexes and Mangrove Units

Gulf of Mexico
Alvarado — Mexico
Usumacinta — Mexico

Yucatán
Petenes — Mexico
Río Lagartos — Mexico
Mayan Corridor — Mexico
Belizean Coast — Mexico, Belize, Guatemala
Belizean Reef — Mexico, Belize

Atlantic Central America
Northern Honduras—Honduras
Mosquitia/Nicaraguan Caribbean
Coast—Honduras, Nicaragua
Río Negro/Río San Sun—Nicaragua,
Costa Rica
Bocas del Toro/Bastimentos Island/San
Blas—Panama

West Indies
Bahamas—Bahamas, Turks & Caicos
Islands, U.S.
Greater Antilles—Cuba, Jamaica,
Hispaniola, Puerto Rico, Cayman
Islands
Lesser Antilles—Lesser Antilles

Continental Caribbean
Coastal Venezuela—Venezuela,
Colombia, Netherlands Antilles
Magdalena/Santa Marta—
Colombia

Amazon-Orinoco-Maranhão
Trinidad—Trinidad & Tobago
Segment 0: Orinoco-Cabo Orange—
Venezuela, Guyana, French Guiana
Segment I: Cabo Orange-Cabo Norte—
Brazil
Segment II: Cabo Norte-Ponta Curuçá—
Brazil
Segment III: Ponta Curuçá-Parnaíba
Delta—Brazil

NE Brazil
Segment IV: Parnaíba Delta-Cabo
Calcanhar—Brazil
Segment V: Cabo Calcanhar-Recôncavo
Baiano—Brazil

Segment VI: Recôncavo Baiano-Cabo
Frio—Brazil

SE Brazil
Segment VII: Cabo Frio-Laguna—Brazil

Pacific Mangrove Complexes and Mangrove Units

Sea of Cortez
NW Mexican Coast—Mexico
Marismas Nacionales/San Blas—Mexico

Southern Mexico
Southern Pacific Coast of Mexico—Mexico

Pacific Central America
Tehuantepec/El Manchón—Mexico,
Guatemala
Northern Dry Pacific Coast—Guatemala,
El Salvador
Gulf of Fonseca—El Salvador, Honduras,
Nicaragua
Southern Dry Pacific Coast—Nicaragua,
Costa Rica
Moist Pacific Coast—Costa Rica, Panama
Panama Dry Pacific Coast—Panama

Pacific South America
Gulf of Panama—Panama, Colombia
Esmeraldas/Pacific Colombia—
Colombia, Ecuador
Manabí—Ecuador
Gulf of Guayaquil/Tumbes—Ecuador,
Peru
Piura—Peru

Galapagos
Galapagos Islands—Ecuador

Appendix E

Results of Assessments of Landscape-Level Criteria, Conservation Status, and Biological Distinctiveness of Non-Mangrove Ecoregions

MAJOR ECOSYSTEM TYPE / Major Habitat Type / Bioregion / Ecoregion	Ecoregion Number	Ecoregion Size (km²)	Habitat Loss a	Habitat Blocks b	Fragmentation c	Degradation d	Conversion e	Protection f	Total g	Snapshot Conservation Status h	Final Conservation Status i	Biological Distinctiveness j	Biodiversity Priority k
TROPICAL BROADLEAF FORESTS													
Tropical Moist Broadleaf Forests													
Caribbean Tropical Moist Forests													
Cuban moist forests—Cuba	1	20,069	32	10	8	n.a.	6	4	60	3	3	2	I
Hispaniolan moist forests—Haiti, Dominican Republic	2	43,136	40	10	12	n.a.	8	4	74	2	2	2	I
Jamaican moist forests—Jamaica	3	7,849	40	15	16	n.a.	8	10	89	2	2	2	I
Puerto Rican moist forests—Puerto Rico	4	7,237	32	15	12	n.a.	0	6	65	2	3	3	III
Windward Islands moist forests—Windward Islands	5	1,914	20	5	0	n.a.	8	8	41	3	3	3	III
Leeward Islands moist forests—Leeward Islands	6	951	10	5	0	n.a.	0	1	16	4	4	3	III
Central American Tropical Moist Forests													
Oaxacan moist forests—Mexico	7	4,715	—	—	—	n.a.	—	—	—	2	2	3	II
Tehuantepec moist forests—Mexico, Guatemala	8	146,752	20	5	16	n.a.	6	6	53	3	2	3	II*
Yucatán moist forests—Mexico	9	64,012	20	2	0	n.a.	6	4	32	4	3	3	III
Sierra Madre moist forests—Mexico, Guatemala, El Salvador	10	9,137	32	20	5	n.a.	9	8	74	2	2	3	II

Note: Footnotes explaining the symbols and numerical codes in this table are found at the end of the appendix.

MAJOR ECOSYSTEM TYPE
Major Habitat Type

Bioregion / Ecoregion	Ecoregion Number	Ecoregion Size (km²)	Habitat Loss [a]	Habitat Blocks [b]	Fragmentation [c]	Degradation [d]	Conversion [e]	Protection [f]	Total [g]	Snapshot Conservation Status [h]	Final Conservation Status [i]	Biological Distinctiveness [j]	Biodiversity Priority [k]
Central American Tropical Moist Forests (continued)													
Central American montane forests—Mexico, El Salvador, Guatemala, Honduras	11	7,676	20	15	16	n.a.	6	6	63	3	2	3	II
Belizean swamp forests—Belize	12	4,150	32	15	12	n.a.	6	10	75	2	2	4	III
Central American Atlantic moist forests—Guatemala, Belize Honduras, Nicaragua, Costa Rica, Panama	13	155,020	32	2	12	n.a.	10	1	57	3	3	3	III
Costa Rican seasonal moist forests—Costa Rica, Nicaragua	14	10,620	40	20	20	n.a.	8	10	98	1	1	4	III
Isthmian-Pacific moist forests—Costa Rica, Panama	15	28,881	32	15	16	n.a.	6	6	75	2	2	3	II
Talamancan montane forests—Costa Rica, Panama	16	15,991	20	5	0	n.a.	6	1	32	4	4	2	II
Orinoco Tropical Moist Forests													
Cordillera La Costa montane forests—Venezuela	17	13,481	32	2	12	n.a.	0	1	47	3	3	2	I
Orinoco Delta swamp forests—Venezuela, Guyana	18	31,698	0	2	0	n.a.	0	1	3	5	4	3	III
Trinidad & Tobago moist forests—Trinidad & Tobago	19	4,456	20	2	12	n.a.	6	4	44	3	3	4	IV
Guianan highlands moist forests—Venezuela, Brazil, Guyana	20	248,018	—	—	—	n.a.	—	—	—	5	5	2	III
Tepuis—Venezuela, Brazil, Guyana, Suriname, Colombia	21	49,157	0	2	0	n.a.	0	1	3	5	5	1	II
Napo moist forests—Peru, Ecuador, Colombia	22	369,847	0	2	5	n.a.	6	1	14	4	4	1	I
Amazonian Tropical Moist Forests													
Macarena montane forests—Colombia	23	2,366	10	5	5	n.a.	6	10	36	4	3	2	I

No.	Ecoregion	Area											
24	Japura/Negro moist forests—Colombia, Venezuela, Brazil, Peru	718,551	0	2	0	n.a.	0	1	3	5	5	1	II
25	Uatama moist forests—Brazil, Venezuela, Guyana	288,128	0	2	0	n.a.	0	4	6	5	4	3	III
26	Amapá moist forests—Brazil, Suriname	195,120	0	2	0	n.a.	0	4	6	5	4	3	III
27	Guianan moist forests—Venezuela, Guyana, Suriname, Brazil, French Guiana	457,017	0	2	0	n.a.	0	1	3	5	4	3	III
28	Paramaribo swamp forests—Suriname	7,760	10	2	5	n.a.	8	6	31	4	3	3	III
29	Ucayali moist forests—Brazil, Peru	173,527	10	2	0	n.a.	6	4	22	4	3	1	I
30	Western Amazonian swamp forests—Peru, Colombia	8,315	—	—	—	n.a.	—	—	—	4	4	1	I
31	Southwestern Amazonian moist forests—Brazil, Peru, Bolivia	534,316	0	2	0	n.a.	0	1	3	5	4	1	I
32	Juruá moist forests—Brazil	361,055	0	2	0	n.a.	0	1	3	5	5	2	III
33	Varzea forests—Brazil, Peru, Colombia	193,129	20	2	12	n.a.	6	4	44	3	3	1	I
34	Purus/Madeira moist forests—Brazil	561,765	0	2	0	n.a.	0	6	8	4	4	4	IV
35	Rondônia/Mato Grosso moist forests—Brazil, Bolivia	645,089	10	2	5	n.a.	6	4	27	4	3	2	II
36	Beni swamp and gallery forests—Bolivia	31,329	0	2	0	n.a.	0	4	6	5	4	4	IV
37	Tapajós/Xingu moist forests—Brazil	630,905	10	2	5	n.a.	6	6	29	4	3	4	IV
38	Tocantins moist forests—Brazil	279,419	32	5	16	n.a.	6	8	67	2	2	4	III
	Northern Andean Tropical Moist Forests												
39	Chocó/Darién moist forests—Colombia, Panama, Ecuador	82,079	10	2	0	n.a.	9	8	29	4	3	1	I
40	Eastern Panamanian montane forests—Panama, Colombia	2,905	0	10	12	n.a.	6	8	36	4	3	3	III
41	Northwestern Andean montane forests—Colombia, Ecuador	52,937	20	10	12	n.a.	6	6	54	3	2	1	I

MAJOR ECOSYSTEM TYPE
Major Habitat Type
Bioregion
Ecoregion

Bioregion / Ecoregion	Ecoregion Number	Ecoregion Size (km²)	Habitat Loss [a]	Habitat Blocks [b]	Fragmentation [c]	Degradation [d]	Conversion [e]	Protection [f]	Total [g]	Snapshot Conservation Status [h]	Final Conservation Status [i]	Biological Distinctiveness [j]	Biodiversity Priority [k]
Northern Andean Tropical Moist Forests (continued)													
Western Ecuador moist forests—Ecuador, Colombia	42	40,218	40	20	12	n.a.	6	8	86	2	1	2	I
Cauca Valley montane forests—Colombia	43	32,412	32	20	20	n.a.	6	10	88	2	1	1	I
Magdalena Valley montane forests—Colombia	44	49,322	32	20	20	n.a.	6	10	88	2	1	1	I
Magdalena/Urabá moist forests—Colombia	45	73,660	32	15	16	n.a.	6	10	79	2	2	3	II
Cordillera Oriental montane forests—Colombia, Venezuela	46	66,712	20	2	12	n.a.	9	4	47	3	3	1	I
Eastern Cordillera Real montane forests—Ecuador, Colombia, Peru	47	84,442	20	2	12	n.a.	8	8	50	3	3	1	I
Santa Marta montane forests—Colombia	48	4,707	32	5	5	n.a.	6	4	52	3	3	2	I
Venezuelan Andes montane forests—Venezuela, Colombia	49	16,638	10	15	16	n.a.	6	4	51	3	2	1	I
Catatumbo moist forests—Venezuela, Colombia	50	21,813	40	20	20	n.a.	0	10	90	1	1	4	III
Central Andean Tropical Moist Forests													
Peruvian Yungas—Peru	51	188,735	20	2	12	n.a.	8	8	50	3	2	1	I
Bolivian Yungas—Bolivia, Argentina	52	72,517	20	2	12	n.a.	8	8	50	3	2	2	I
Andean Yungas—Argentina, Bolivia	53	55,457	20	2	12	n.a.	6	1	41	3	3	3	III
Eastern South American Tropical Moist Forests													
Brazilian Coastal Atlantic forests—Brazil	54	233,266	40	15	20	n.a.	10	6	91	1	1	1	I
Brazilian Interior Atlantic forests—Brazil, Argentina, Paraguay	55	803,908	32	5	20	n.a.	8	7	72	2	2	2	I

Tropical Dry Broadleaf Forests

Caribbean Tropical Dry Forests

56	Cuban dry forests—Cuba	61,466	40	10	16	n.a.	6	8	80	2	2	3	II*
57	Hispaniolan dry forests—Haiti, Dominican Republic	14,610	32	5	10	n.a.	6	4	57	3	2	3	II
58	Jamaican dry forests—Jamaica	2,189	20	20	5	n.a.	10	10	65	2	2	4	III
59	Puerto Rican dry forests—Puerto Rico	1,295	32	20	16	n.a.	6	10	84	2	2	4	III
60	Bahamian dry forests—Bahamas, Turks & Caicos Islands	9,293	20	5	5	n.a.	8	8	46	4	3	4	IV
61	Cayman Islands dry forests—Cayman Islands	230	—	—	—	n.a.	—	—	—	—	2	4	III
62	Windward Islands dry forests—Windward Islands	431	32	10	5	n.a.	6	8	61	3	2	4	III
63	Leeward Islands dry forests—Leeward Islands	182	40	15	20	n.a.	6	8	89	1	1	4	III

Northern Mexican Tropical Dry Forests

64	Baja California dry forests—Mexico	93	0	2	0	n.a.	0	10	12	4	4	4	IV
65	Sinaloan dry forests—Mexico	119,184	20	5	5	n.a.	8	10	48	3	3	3	III
66	Tamaulipas/Veracruz dry forests—Mexico	40,858	32	15	16	n.a.	6	10	79	2	2	4	III*

Central American Tropical Dry Forests

67	Jalisco dry forests—Mexico	19,973	32	5	12	n.a.	9	8	66	2	2	2	I
68	Balsas dry forests—Mexico	161,098	20	2	12	n.a.	8	10	52	3	2	2	I
69	Oaxacan dry forests—Mexico	10,566	32	15	16	n.a.	6	10	79	2	2	3	II
70	Veracruz dry forests—Mexico	35,546	40	20	20	n.a.	3	10	93	1	1	4	III
71	Yucatán dry forests—Mexico	45,554	32	15	16	n.a.	9	10	82	2	2	4	III
72	Central American Pacific dry forests—El Salvador, Honduras, Nicaragua, Costa Rica, Guatemala	50,101	40	20	20	n.a.	6	4	90	1	1	3	II
73	Panamanian dry forests—Panama	5,010	40	20	20	n.a.	6	8	94	1	1	4	III

MAJOR ECOSYSTEM TYPE
Major Habitat Type
Bioregion
Ecoregion

Ecoregion	Ecoregion Number	Ecoregion Size (km²)	Habitat Loss [a]	Habitat Blocks [b]	Fragmentation [c]	Degradation [d]	Conversion [e]	Protection [f]	Total [g]	Snapshot Conservation Status [h]	Final Conservation Status [i]	Biological Distinctiveness [j]	Biodiversity Priority [k]
Orinoco Tropical Dry Forests													
Llanos dry forests—Venezuela	74	44,177	32	15	16	n.a.	6	10	79	2	2	4	III*
Trinidad & Tobago dry forests—Trinidad & Tobago	75	251	32	15	12	n.a.	6	8	73	2	2	4	III
Amazonian Tropical Dry Forests													
Bolivian lowland dry forests—Bolivia, Brazil	76	156,814	40	15	20	n.a.	0	10	85	2	1	1	I
Northern Andean Tropical Dry Forests													
Cauca Valley dry forests—Colombia	77	5,130	40	20	20	n.a.	6	10	96	1	1	4	III
Magdalena Valley dry forests—Colombia	78	13,837	40	20	20	n.a.	6	10	96	1	1	4	III
Patía Valley dry forests—Colombia	79	1,291	40	20	20	n.a.	6	10	96	1	1	4	III
Sinú Valley dry forests—Colombia	80	55,473	40	20	20	n.a.	6	10	96	1	1	4	III
Ecuadorian dry forests—Ecuador	81	22,271	40	20	20	n.a.	6	10	96	1	1	1	I
Tumbes/Piura dry forests—Ecuador, Peru	82	64,588	32	15	12	n.a.	6	6	71	2	2	1	I
Marañón dry forests—Peru	83	14,921	—	—	—	n.a.	—	—	—	2	2	3	II
Maracaibo dry forests—Venezuela	84	31,471	32	15	12	n.a.	6	8	73	2	2	4	III
Lara/Falcón dry forests—Venezuela	85	16,178	32	15	12	n.a.	6	8	73	2	2	4	III
Central Andean Tropical Dry Forests													
Bolivian montane dry forests—Bolivia	86	39,368	40	20	20	n.a.	6	10	96	1	1	3	II*

CONIFER/TEMPERATE BROADLEAF FORESTS

Temperate Forests

Southern South American Temperate Forests

Ecoregion													
Chilean winter-rain forests—Chile	87	24,937	32	15	18	n.a.	6	8	79	2	2	2	I
Valdivian temperate forests—Chile, Argentina	88	166,248	10	10	12	n.a.	8	1	41	3	3	1	I
Subpolar *Nothofagus* forests—Chile, Argentina	89	141,120	0	2	12	n.a.	6	1	21	4	3	3	III

Tropical and Subtropical Coniferous Forests

Caribbean Tropical and Subtropical Coniferous Forests

Ecoregion													
Cuban pine forests—Cuba	90	6,017	32	10	5	n.a.	3	8	58	3	3	2	I
Hispaniolan pine forests—Haiti, Dominican Republic	91	10,833	32	5	12	n.a.	6	4	59	3	3	2	I
Bahamian pine forests—Bahamas, Turks & Caicos Islands	92	3,920	32	10	0	n.a.	0	10	52	3	3	4	IV

Northern Mexican Tropical and Subtropical Coniferous Forests

Ecoregion													
Sierra Juarez pine-oak forests—Mexico, U.S.	93	24,229	20	2	5	n.a.	6	8	41	3	3	4	IV
San Lucan pine-oak forests—Mexico, U.S.	94	895	0	2	0	n.a.	0	1	3	5	5	4	IV
Sierra Madre Occidental pine-oak forests—Mexico, U.S.	95	204,374	32	15	12	n.a.	9	10	78	2	2	1	I
Central Mexican pine-oak forests—Mexico	96	3,719	32	10	20	n.a.	0	10	72	2	2	3	II
Sierra Madre Oriental pine-oak forests—Mexico, U.S.	97	38,200	10	2	0	n.a.	8	6	26	4	4	2	II
Veracruz pine-oak forests—Mexico	98	5,979	—	—	—	n.a.	—	—	—	2	2	1	III

Central American Tropical and Subtropical Coniferous Forests

Ecoregion													
Mexican transvolcanic pine-oak forests—Mexico	99	72,802	32	20	16	n.a.	8	10	86	2	2	2	I
Veracruz montane forests—Mexico	100	6,615	0	2	0	n.a.	8	6	16	4	4	3	III
Sierra Madre del Sur pine-oak forests—Mexico	101	41,129	—	—	—	n.a.	—	—	—	2	1	1	I

MAJOR ECOSYSTEM TYPE
Major Habitat Type
Bioregion
Ecoregion

Ecoregion	Ecoregion Number	Ecoregion Size (km²)	Habitat Loss [a]	Habitat Blocks [b]	Fragmentation [c]	Degradation [d]	Conversion [e]	Protection [f]	Total [g]	Snapshot Conservation Status [h]	Final Conservation Status [i]	Biological Distinctiveness [j]	Biodiversity Priority [k]
Central American Tropical and Subtropical Coniferous Forests (continued)													
Central American pine-oak forests—Guatemala, El Salvador, Honduras, Mexico, Nicaragua	102	127,910	32	2	12	n.a.	9	1	56	3	3	3	III
Belizean pine forests—Belize	103	2,750	—	—	—	n.a.	—	—	—	4	4	2	II
Miskito pine forests—Nicaragua, Honduras	104	15,064	10	5	5	n.a.	6	10	36	4	4	2	II
Eastern South American Tropical and Subtropical Coniferous Forests													
Brazilian Araucaria forests—Brazil, Argentina	105	206,459	40	20	20	n.a.	10	10	100	1	1	3	II*

GRASSLANDS/SAVANNAS/SHRUBLANDS
Grasslands, Savannas, and Shrublands

Ecoregion	Ecoregion Number	Ecoregion Size (km²)	Habitat Loss [a]	Habitat Blocks [b]	Fragmentation [c]	Degradation [d]	Conversion [e]	Protection [f]	Total [g]	Snapshot Conservation Status [h]	Final Conservation Status [i]	Biological Distinctiveness [j]	Biodiversity Priority [k]
Northern Mexican Grasslands, Savannas, and Shrublands													
Central Mexican grasslands—Mexico, U.S.	106	199,919	—	—	—	—	—	—	—	—	—	3	—
Eastern Mexican grasslands—Mexico	107	3,674	—	—	—	—	—	—	—	—	—	4	—
Central American Grasslands, Savannas, and Shrublands													
Tabasco/Veracruz savannas—Mexico	108	9,252	—	—	—	—	—	—	—	1	1	4	III*
Tehuantepec savannas—Mexico	109	5,614	—	—	—	—	—	—	—	1	1	4	III
Orinoco Grasslands, Savannas, and Shrublands													
Llanos—Venezuela, Colombia	110	355,112	20	2	0	5	6	1	34	4	4	3	III
Amazonian Grasslands, Savannas, and Shrublands													
Guianan savannas—Suriname, Guyana, Brazil, Venezuela	111	128,375	10	2	4	0	6	6	28	4	4	3	III
Amazonian savannas—Brazil, Colombia, Venezuela	112	120,124	10	2	4	0	6	6	28	4	4	3	III
Beni savannas—Bolivia	113	165,445	32	2	8	0	0	8	50	3	2	3	II*

No.	Ecoregion												
	Eastern South American Grasslands, Savannas, and Shrublands												
114	Cerrado—Brazil, Paraguay, Bolivia	1,982,249	20	2	8	7	8	1	46	3	3	1	I
115	Chaco savannas—Argentina, Paraguay, Bolivia, Brazil	611,053	20	2	8	3	8	4	45	3	3	2	I
116	Humid Chaco—Argentina, Paraguay, Uruguay, Brazil	474,340	10	2	8	3	6	8	37	3	3	4	IV
117	Córdoba montane savannas—Argentina	55,798	10	2	6	0	4	10	32	4	3	4	IV
	Southern South American Grasslands, Savannas, and Shrublands												
118	Argentine Monte—Argentina	197,710	10	2	8	0	6	6	32	4	4	3	III
119	Argentine Espinal—Argentina	207,054	10	2	8	6	8	6	40	3	3	3	III
120	Pampas—Argentina	426,577	40	15	8	8	6	8	85	2	2	3	II*
121	Uruguayan savannas—Uruguay, Brazil, Argentina	336,846	20	15	5	6	6	8	60	3	3	3	III
	Flooded Grasslands												
	Caribbean Flooded Grasslands												
122	Cuban wetlands—Cuba	5,345	—	—	—	—	—	—	—	—	2	2	I
123	Enriquillo wetlands—Haiti, Dominican Republic	574	—	—	—	—	—	—	—	—	3	2	I
	Northern Mexican Flooded Grasslands												
124	Central Mexican wetlands—Mexico	362	—	—	—	—	—	—	—	—	1	2	I
	Central American Flooded Grasslands												
125	Jalisco palm savannas—Mexico	555	—	—	—	—	—	—	—	1	1	3	II*
126	Veracruz palm savannas—Mexico	7,518	—	—	—	—	—	—	—	1	1	3	II
127	Quintana Roo wetlands—Mexico	2,452	0	2	0	0	1	3	—	5	5	3	IV
	Orinoco Flooded Grasslands												
128	Orinoco wetlands—Venezuela	6,403	—	—	—	—	—	—	—	4	4	3	III

MAJOR ECOSYSTEM TYPE
Major Habitat Type

Bioregion Ecoregion	Ecoregion Number	Ecoregion Size (km²)	Habitat Loss [a]	Habitat Blocks [b]	Fragmentation [c]	Degradation [d]	Conversion [e]	Protection [f]	Total [g]	Snapshot Conservation Status [h]	Final Conservation Status [i]	Biological Distinctiveness [j]	Biodiversity Priority [k]
Amazonian Flooded Grasslands													
Western Amazonian flooded grasslands—Peru, Bolivia	129	10,111	—	—	—	—	—	—	—	4	4	3	III
Eastern Amazonian flooded grasslands—Brazil	130	69,533	—	—	—	—	—	—	—	3	3	3	III*
São Luis flooded grasslands—Brazil	131	1,681	—	—	—	—	—	—	—	2	2	4	III
Northern Andean Flooded Grasslands													
Guayaquil flooded grasslands—Ecuador	132	3,617	—	—	—	—	—	—	—	2	2	3	II*
Eastern South American Flooded Grasslands													
Pantanal—Brazil, Bolivia, Paraguay	133	140,927	10	2	0	6	10	7	35	4	3	1	I
Paraná flooded savannas—Argentina	134	36,452	—	—	5	—	—	—	—	2	2	3	II
Montane Grasslands													
Central American Montane Grasslands													
Mexican alpine tundra—Mexico	135	147	20	5	12	n.a.	6	4	47	3	3	3	III*
Costa Rican paramo—Costa Rica	136	31	10	2	5	n.a.	0	1	18	4	4	3	III
Northern Andean Montane Grasslands													
Santa Marta paramo—Colombia	137	1,329	—	—	—	n.a.	—	—	—	—	3	1	I
Cordillera de Mérida paramo—Venezuela	138	3,518	0	2	16	n.a.	0	1	19	4	4	1	I
Northern Andean paramo—Colombia, Ecuador	139	58,806	0	10	16	n.a.	6	4	36	4	3	1	I
Central Andean Montane Grasslands													
Cordillera Central paramo—Peru, Ecuador	140	14,128	0	10	16	n.a.	0	10	36	4	3	1	I

No.	Ecoregion	Area											BD
141	Central Andean puna—Bolivia, Argentina, Peru, Chile	183,868	20	5	6	n.a.	6	4	41	3	3	2	I
142	Central Andean wet puna—Peru, Bolivia, Chile	188,911	20	2	8	n.a.	6	1	37	3	3	2	I
143	Central Andean dry puna—Argentina, Bolivia, Chile	232,958	10	2	4	n.a.	6	10	32	4	3	2	I
Southern South American Montane Grasslands													
144	Southern Andean steppe—Argentina, Chile	198,643	10	2	0	n.a.	0	4	16	4	4	4	IV
145	Patagonian steppe—Argentina, Chile	474,757	32	2	0	n.a.	2	6	42	3	3	2	I
146	Patagonian grasslands—Argentina, Chile	59,585	32	5	0	n.a.	2	6	45	3	3	3	III
XERIC FORMATIONS													
Mediterranean Scrub													
Northern Mexican Mediterranean Scrub													
147	California coastal sage-chaparral—Mexico, U.S.	27,104	—	—	—	n.a.	—	—	—	2	1	1	I
Central Andean Mediterranean Scrub													
148	Chilean matorral—Chile	141,643	—	—	—	n.a.	—	—	—	2	2	1	I
Deserts and Xeric Shrublands													
Caribbean Deserts and Xeric Shrublands													
149	Cuban cactus scrub—Cuba	3,044	20	10	5	n.a.	3	8	46	3	3	4	IV
150	Cayman Islands xeric scrub—Cayman Islands	32	32	15	5	n.a.	8	8	68	2	2	4	III
151	Windward Islands xeric scrub—Windward Islands	944	32	15	5	n.a.	6	10	68	2	2	4	III
152	Leeward Islands xeric scrub—Leeward Islands	1,521	40	20	20	n.a.	6	10	96	1	1	4	III*
Northern Mexican Deserts and Xeric Shrublands													
153	Baja California xeric scrub—Mexico	72,377	0	2	0	n.a.	0	1	3	5	5	3	IV
154	San Lucan mezquital—Mexico	2,226	—	—	—	n.a.	—	—	—	5	5	4	IV
155	Western Mexican mezquital—Mexico, U.S.	22,894	—	—	—	n.a.	—	—	—	—	4	—	—

MAJOR ECOSYSTEM TYPE
Major Habitat Type
Bioregion

Ecoregion	Ecoregion Number	Ecoregion Size (km²)	Habitat Loss a	Habitat Blocks b	Fragmentation c	Degradation d	Conversion e	Protection f	Total g	Snapshot Conservation Status h	Final Conservation Status i	Biological Distinctiveness i	Biodiversity Priority k
Northern Mexican Deserts and Xeric Shrublands (continued)													
Sonoran xeric scrub—Mexico, U.S.	156	232,340	20	2	5	n.a.	8	1	36	4	4	2	II
Northern Sonoran cactus scrub—Mexico, U.S.	157	97,962	10	5	5	n.a.	0	10	30	4	4	1	I
Mexican Interior chaparral—Mexico, U.S.	158	22,252	—	—	—	n.a.	—	—	—	—	—	4	—
Chihuahuan xeric scrub—Mexico, U.S.	159	399,446	20	2	12	n.a.	8	8	50	3	3	4	IV
Central Mexican mezquital—Mexico	160	29,347	32	15	20	n.a.	6	8	81	2	2	4	III
Eastern Mexican matorral—Mexico	161	26,684	—	—	—	n.a.	—	—	—	—	—	3	—
Eastern Mexican mezquital—Mexico, U.S.	162	138,696	32	5	5	n.a.	8	10	60	3	3	4	IV
Central Mexican cactus scrub—Mexico	163	37,860	—	—	—	n.a.	—	—	—	—	—	3	—
Central American Deserts and Xeric Shrublands													
Pueblan xeric scrub—Mexico	164	6,818	32	20	16	n.a.	6	10	84	1	1	3	II*
Guerreran cactus scrub—Mexico	165	5,232	—	—	—	n.a.	—	—	—	—	3	3	III
Motagua Valley thornscrub—Guatemala	166	2,363	40	20	20	n.a.	8	10	98	1	1	3	II
Orinoco Deserts and Xeric Shrublands													
Aruba/Curaçao/Bonaire cactus scrub—Netherlands Antilles	167	407	32	10	5	n.a.	0	8	55	3	3	4	IV
La Costa xeric shrublands—Venezuela	168	64,379	32	15	12	n.a.	6	8	73	2	2	4	III
Araya and Paría xeric scrub—Venezuela	169	5,424	32	15	12	n.a.	6	8	73	2	2	3	II*

	Ecoregion	Area	a	b	c	d	e	f	g	h	i	j	k
Northern Andean Deserts and Xeric Shrublands													
170	Galapagos Islands xeric scrub – Ecuador	9,122	10	2	5	n.a.	0	1	18	4	3	1	I
171	Guajira/Barranquilla xeric scrub – Colombia, Venezuela	32,404	32	15	12	n.a.	6	8	73	2	2	3	II
172	Paraguaná xeric scrub – Venezuela	15,987	32	15	12	n.a.	6	8	73	2	2	3	II
Central Andean Deserts and Xeric Shrublands													
173	Sechura desert – Peru, Chile	189,928	10	10	5	n.a.	6	8	39	3	3	3	III*
174	Atacama desert – Chile	103,841	–	–	–	n.a.	–	–	–	3	3	3	III
Eastern South American Deserts and Xeric Shrublands													
175	Caatinga – Brazil	752,606	32	2	0	n.a.	6	8	48	3	3	3	III*
Restingas													
Northern Andean Restingas													
176	Paraguaná restingas – Venezuela	15,987	32	15	12	n.a.	6	8	73	2	2	3	II*
Amazonian Restingas													
177	Northeastern Brazil restingas – Brazil	10,248	40	20	20	n.a.	10	10	100	1	1	1	I
Eastern South American Restingas													
178	Brazilian Atlantic Coast restingas – Brazil	8,740	40	20	16	n.a.	10	10	96	1	1	1	I

Explanation of Numerical Codes (see Appendix A for details)

a. Habitat loss: Index from 0 (least loss) to 40 (most).

b. Habitat blocks: Index from 2 (largest and most numerous blocks) to 20 (smallest and least numerous).

c. Fragmentation: Index from 0 (least fragmentation) to 20 (most); index ranges from 0 to 10 if degradation value present.

d. Degradation: Index from 0 (least degraded) to 10 (most) used for two MHTs – grasslands, savannas, and shrublands, and flooded grasslands (otherwise indicated as n.a. – not applicable).

e. Conversion: Index from 0 (lowest annual conversion rate of natural habitat) to 10 (highest rate).

f. Protection: Index from 1 (best protection) to 10 (least).

g. Total: Index for snapshot conservation status ranging from 0 (best) to 100 (worst).

h. Snapshot conservation status: 1=Critical (89-100 points), 2=Endangered (65-88), 3=Vulnerable (37-64), 4=Relatively Stable (7-36), and 5=Relatively Intact (0-6).

i. Final conservation status: Snapshot status modified by threat assessment; codes are the same as above.

j. Biological distinctiveness: 1=Globally Outstanding, 2=Regionally Outstanding, 3= Bioregionally Outstanding, and 4=Locally Important.

k. Biodiversity priority: I=Highest Priority at Regional Scale, II=High Priority at Regional Scale, III=Moderate Priority at Regional Scale, and IV=Important at National Scale.

* Indicates an ecoregion elevated to a level I[a] priority ranking to achieve better bioregional representation.

– Value is missing because sufficient high quality data were not available.

Appendix F

Ecoregion Profiles and Sources Consulted for Their Delineation, Classification, and Assessment

Ecoregion Delineation and Classification

The ecoregion maps developed for this study classify all terrestrial ecoregions of LAC into standardized major ecosystem and major habitat type categories to facilitate continental-scale analyses and comparisons at biogeographic scales appropriate for conservation. Development of the ecoregion maps was challenging because the quality of ecological information available varied widely throughout the LAC region and no standardized system existed at the desired level of biogeographic resolution. In order to facilitate future refinements, we highlight here some particularly challenging issues and areas.

Currently, there is no clear consensus on ecoregion delineation in *Amazonia and the Guianas*. Geographic patterns of biodiversity in this vast region are not well known. However, there is evidence of significant heterogeneity over a wide range of biogeographic scales. For this reason, the large ecoregion units used in this study for Amazonia should be subdivided in future studies to more accurately reflect patterns of biodiversity and better address conservation needs.

The ecoregion units generally follow divisions proposed on phytogeographers' vegetation maps (e.g., Ducke and Black 1953; Rizzini 1963; Hueck 1966; Prance 1973, 1977; Fundação IBGE 1993)[1] for the Amazon, and priority areas proposed on the conservation maps developed for Amazonia at the international workshop "Biological Priorities for Conservation in Amazonia" in Manaus, Brazil in January, 1990 (Conservation International 1990; Rylands 1990;

in delineating Amazonian ecoregions, we attempted to avoid splitting the areas identified at the Manaus Workshop as highest priority for conservation based on biodiversity and endemism). Our delineations also incorporate information from a series of preliminary maps of ecological units for Amazonia and South America developed by Luis Diego Gómez for The Nature Conservancy. These maps divide the region on the basis of biophysical parameters and patterns of plant diversity.

Clinebell et al.'s (1995) analysis of soil and climatic data as predictors for neotropical tree and liana species richness represents an important advance in understanding how biophysical parameters can be used for mapping biodiversity. Building on the work of Diego Gómez, Clinebell et al., CIAT/EMBRAPA-CPAC (1985), and others, further analyses of biophysical parameters such as soil type, rainfall, and seasonality should be conducted for Amazonia and other regions of LAC to better predict patterns of species richness, endemism, and vegetation types, and to improve ecoregion maps. The conservation analysis of Amazonian watersheds by Peres and Terborgh (1995) provides a new perspective on the factors that contribute to effective conservation units. Future analyses should look closely at the conservation benefits of using watershed boundaries and other defensibility considerations for delineating conservation units.

Tepui formations occur from Suriname westward to the foothills of the Andes. Many of the more westerly lower elevation formations are not well mapped (Brown 1987) and do not appear on our maps because of the coarse scale. Huber and Alarcon's (1988) map of vegetation types of Venezuela provides excellent information on tepuis and other vegetative formations in that country.

---

1. Sources cited in Appendix F are listed on pages 106-116, rather than in the general References section at the back of this book.

We had difficulty in locating detailed or complete ecological maps for *Nicaragua, Honduras, El Salvador, Guatemala, Uruguay, Chile, Dominican Republic, Haiti, and several Caribbean Islands*. The ecoregion maps used in this study would benefit from revisions and updates in these areas.

Detailed ecological maps of *Peru* by G. Lamas in Prance (1982) could be compared with the ecoregion units used in this study to refine boundaries and classifications for that region.

Several *Mexican xeric and grassland ecoregions* were not fully analyzed in this study and many of these complex systems would benefit from further review of their classifications and boundaries.

We recognize that many ecoregions have *strong ecological gradients* or contain a wide variety of habitat types. For example, extensive tracts of dry forests occur in both the Cerrado and western portions of the Guianan moist forests. The conservation needs of these distinct habitat types should be addressed at the sub-ecoregional level.

Small *oceanic islands* were generally not analyzed in this study. Some support endemic species and distinct natural communities and are highly threatened. We recommend a conservation analysis of islands be conducted. Examples of islands not analyzed include: San Andrés and Providencia Archipelago; Malpelo; Fernando de Noronha; Juan Fernández Archipelago; Desventuradas; Cocos; Trindade and Martin Vaz Islands; Revillagigedo Archipelago; Marias Islands; and Guadalupe Island. Some offshore islands were specifically included in adjacent mainland ecoregions: Coiba and Perlas Islands of Panama, islands in the Gulf of California, and Margarita and Tortuga Islands near Venezuela. The Malvinas (Falkland Islands) are included in the Patagonian grasslands[2] but they may merit being delineated as a separate ecoregion after further review of their biogeography. The Galapagos Islands were considered sufficiently distinct to be delineated as a separate ecoregion.

Four critical *landscape-level issues* need to be addressed more fully in future conservation assessments: (a) ecological linkages among adjacent ecoregions (e.g., seasonal or altitudinal migrations; movements due to large-scale disturbances; gene flow and metapopulation interactions for widespread species; downstream effects of ecological changes in upper watersheds; linkages between terrestrial and aquatic ecosystems); (b) critical habitats and corridors for continental-scale migrations of vertebrates and invertebrates; (c) how best to conserve large blocks of intact habitat that encompass portions of several ecoregions and regions with high levels of ecoregion complexity (i.e., beta and gamma diversity); and (d) the hypothesis that large-scale physical or ecological processes (e.g., rainfall and nutrient cycling) are significantly altered due to habitat loss long before changes in species or communities are evident.

Workshop participants recognized an urgent need to conduct similar priority-setting analyses for *freshwater* (i.e., marshes, rivers, and lakes) and *marine* ecosystems. In this study we delineated as ecoregions only particularly large distinct freshwater complexes such as the Pantanal or the Enriquillo wetlands in Hispaniola. Several freshwater biologists at the BSP workshop developed a preliminary map of freshwater ecoregions and a method for assessing their conservation status. WWF plans to collaborate with Wetlands for the Americas and other conservation organizations and regional experts to build upon this work and help identify conservation priorities for these important ecosystems by developing objective, transparent, and scientific methods to identify conservation priorities among freshwater and marine ecoregions, and conducting a workshop with regional experts to apply these methods to all aquatic ecosystems of Latin America and the Caribbean.

Although patterns of biodiversity and ecological dynamics of freshwater and marine ecosystems are significantly different from those of terrestrial ecosystems, we believe that the integration of information on the conservation status (i.e., the current state or ecological integrity of an ecoregion modified by projected threats) and biological distinctiveness of ecoregions represents an ecologically-sound approach for identifying freshwater and marine priorities. However, the particular set of criteria used to assess conservation status and biological importance must be tailored to the specific ecological requirements, dynamics, patterns of diversity, and responses to disturbance of freshwater and marine ecosystems. Given the substantial differences between marine and freshwater ecosystems, separate methods would be appropriate for each. Appropriate ecoregion classifications and maps for LAC freshwater and marine ecosystems must also be developed.

Ecoregion Profiles

The ecoregion profiles provide the following information:

- The number of the ecoregion, the name of the ecoregion, and the countries in which it occurs.

---

2. A dispute concerning sovereignity over the islands exists between Argentina, which claims this sovereignity, and the U.K., which administers the islands.

- The final conservation status, the biological distinctiveness and the biodiversity conservation priority.
- Size in km² of the ecoregion as calculated from GIS measurements of our ecoregion maps. The values should be interpreted as approximate.
- Sources used for delineation of the ecoregion, for classification of the major habitat type, and for information used to assess the conservation status. References are coded with numbers and are listed at the end of the appendix. References in bold type are the principal references used to define and delineate the ecoregion.
- Notable biodiversity features of Globally and Regionally Outstanding ecoregions and, in many cases, biodiversity summaries for Bioregionally Outstanding and Locally Important ecoregions.
- Major threats to ecoregions as conveyed by regional experts.

Further notes on threat information: The major threats to ecoregions summarized in the profiles were for the most part conveyed by regional experts during the 1994 BSP and WWF LAC Program workshops. Ecoregion-specific threat information was not obtained for a few ecoregions, particularly for several xeric ecoregions of Mexico. The majority of the threats summarized here represent factors that directly contribute to alteration of natural landscapes and ecosystems. Potential ultimate factors such as human population growth or land tenure are not considered.

It is important to recognize that exploitation of wildlife occurs to varying degrees in all LAC ecoregions. Intensive exploitation, particularly hunting activity fueled by commercial enterprises, can quickly reduce or extirpate populations of target species from a region. Populations of large predators (e.g., cats, canids, giant otters) and herbivores (e.g., tapir, vicuña), primates, birds (e.g., guans, curassows, macaws), and larger aquatic species (e.g., caiman) are often targeted and can be rapidly extirpated from habitats, with profound and long-term effects on the composition and function of native ecosystems (see Dirzo and Miranda 1990; Iñigo-Elias and Ramos 1991; Redford 1992; Terborgh 1992). Conservation efforts must strive to conserve habitats that maintain relatively intact faunas, which are typically located in relatively stable and intact ecoregions or in very large blocks of intact habitat, and to reduce and eliminate ecologically unsustainable commercial exploitation and international trade of wildlife. Many plant and invertebrate species are also threatened through commercial harvesting or collecting (e.g., mahogany, cacti, and swallowtail butterflies).

TROPICAL BROADLEAF FORESTS
Tropical Moist Broadleaf Forests
Caribbean

1. Cuban moist forests—Cuba (Vulnerable; Regionally Outstanding; Highest Priority at Regional Scale)
- 20,069 km²
- Sources: 14, 15, **131**, 146

The moist forests of Cuba, and those of the Greater Antilles in general, maintain exceptionally diverse insular biotas with many regional and island endemic species in a wide range of taxa. Cuba, in particular, has a rich moist forest flora. The Greater Antilles are notable for numerous unusual relict species and higher taxa. Expansion of cacao, coffee, and tobacco production are serious threats in some areas.

2. Hispaniolan moist forests—Haiti, Dominican Republic (Endangered; Regionally Outstanding; Highest Priority at Regional Scale)
- 43,136 km²
- Sources: 89, 121, 129, **203**, 205, 207, 208, 232, 241, **256**, 270, 284

Biodiversity considerations are discussed under Cuban moist forests. Logging and agricultural expansion severely threaten the ecoregion. Firewood gathering, grazing, and exploitative hunting are also threats.

3. Jamaican moist forests—Jamaica (Endangered; Regionally Outstanding; Highest Priority at Regional Scale)
- 7,849 km²
- Sources: 17, 20, **21**, **119**, 190

Jamaican moist forests are notable for exceptionally high levels of endemism in a wide range of taxa. Agricultural expansion, particularly of coffee plantations, threaten the ecoregion. Anthropogenic fires, grazing, and the invasion of exotic grasses contribute to the degradation and conversion of native habitat. Unsustainable collecting of the endangered swallowtail butterfly, *Papilio homerus*, occurs.

4. Puerto Rican moist forests—Puerto Rico (Vulnerable; Bioregionally Outstanding; Moderate Priority at Regional Scale)
- 7,237 km²
- Sources: 20, **92**, **177**, 181

Biodiversity considerations are discussed under Cuban moist forests. Urban development pressures, pollution, grazing, roadbuilding, and exotic invasions threaten the ecoregion. Restoration of moist forest has occured in some areas over the last several decades, but it is unclear to what extent the full complement of biodiversity and ecological processes are presently represented in these habitats.

5. Windward Islands moist forests—Windward Islands (Vulnerable; Bioregionally Outstanding; Moderate Priority at Regional Scale)
- 1,914 km²
- Sources: 12, 13, **33, 41, 42, 44-52, 54-56**, 92, **127, 175, 202, 220,** 284

The moist forests of both the Windward and Leeward Islands are comprised of many disjunct forests on different islands. Because many species are endemic to forests on a single island, conservation strategies should emphasize adequate representation of each island. Urban development pressures, pollution, grazing, roadbuilding, and exotic invasions threaten the ecoregion. The types and intensity of threats vary among different islands.

6. Leeward Islands moist forests—Leeward Islands (Relatively Stable; Bioregionally Outstanding; Moderate Priority at Regional Scale)
- 951 km²
- Sources: **9,** 12, 13, **33-39,** 40, **41, 42, 44-52, 54-56,** 92, **127, 175, 202, 220,** 284

Biodiversity considerations are discussed under Windward Islands moist forests. Urban development pressures, pollution, grazing, roadbuilding, and exotic invasions threaten the ecoregion. The types and intensity of threats vary among different islands.

Central America

7. Oaxacan moist forests—Mexico (Endangered; Bioregionally Outstanding; Highest Priority at Regional Scale)
- 4,715 km²
- Sources: **99,** 236; **R. de la Maza and J. Soberón, pers. comm.**

The biodiversity found in this region of Mexico is noted for high levels of endemism, beta diversity, and habitat complexity. Large-scale agricultural expansion of inland coffee plantations and coastal citrus groves, ranching, and logging threaten the ecoregion.

8. Tehuantepec moist forests—Mexico, Guatemala (Endangered; Bioregionally Outstanding; High Priority at Regional Scale)
- 146,752 km²
- Sources: 93, **99,** 106, 107, 112, 113, **178,** 213, 221, 225, 234, 236

Oil exploration and associated roadbuilding, national security road construction, and expanding

cattle ranching pose threats to this ecoregion. Expansion of citrus, banana, and small-scale agriculture threaten forests in the Maya Mountains of Belize.

9. Yucatán moist forests—Mexico (Vulnerable; Bioregionally Outstanding; Moderate Regional Priority)
- 64,012 km²
- Sources: 30, **99,** 225, 236

Ranching, logging, population growth, and resettlement pressure are low-intensity threats to this ecoregion. Development related to tourism could also result in degradation in some areas.

10. Sierra Madre moist forests—Mexico, Guatemala, El Salvador (Endangered; Bioregionally Outstanding; High Priority at Regional Scale)
- 9,137 km²

Sources: 30, **99,** 112, 113, 120, 125, 258, 275

Habitat conversion at middle elevations for "sun coffee" production and firewood gathering threaten the ecoregion.

11. Central American montane forests—Mexico, Guatemala, El Salvador, Honduras (Endangered; Bioregionally Outstanding; High Priority at Regional Scale)
- 7,676 km²
- Sources: 30, 59, **99,** 112, 120, 125, **133,** 169, 198, 203, 216, 236, 244, 260, 265, 275, 282

Conversion of montane forests occurs largely through agricultural expansion and clearing for pastures.

12. Belizean swamp forests—Belize (Endangered; Locally Important; Moderate Priority at Regional Scale)
- 4,150 km²
- Sources: 30, 112, 113, **123**

Although not mapped, examples of this ecoregion also occur in Mexico and Guatemala. This ecoregion faces moderate degradation threats from recreational use and hunting, and high conversion threats from roadbuilding and agricultural expansion.

13. Central American Atlantic moist forests—Guatemala, Belize, Honduras, Nicaragua, Costa Rica, Panama (Vulnerable; Bioregionally Outstanding; Moderate Priority at Regional Scale)
- 155,020 km²
- Sources: 29, 59, 60, 67, 112, 113, 115, 124, 133, 198, 216, **254,** 259, **261,** 271, **273,** 282

Banana plantation and cattle ranch expansion, logging, clearing, and refugee settlements in Nicaragua all pose severe threats to the ecoregion.

Exploitation of parrots and other wildlife is a further threat. Conversion and degradation, even in designated protected areas, will likely increase in the next decade.

14. Costa Rican seasonal moist forests — Costa Rica, Nicaragua (Critical; Locally Important; Moderate Priority at Regional Scale)
- 10,620 km²
- Sources: 85, 117, 247, 254, **261**

Conversion of forests for agriculture and pasture, anthropogenic fires, and degradation from grazing and wildlife exploitation are high-intensity threats to this ecoregion.

15. Isthmian-Pacific moist forests — Costa Rica, Panama (Endangered; Bioregionally Outstanding; High Priority at Regional Scale)
- 28,881 km²
- Sources: 30, 116, 117, **261, 271**

Logging, mining, burning, and habitat conversion for pasture and agriculture represent extreme and high-intensity threats to this ecoregion. Wildlife exploitation and pollution are additional threats.

16. Talamancan montane forests — Costa Rica, Panama (Relatively Stable; Regionally Outstanding; High Priority at Regional Scale)
- 15,991 km²
- Sources: 30, 117, 247, **261, 271**

The Talamancan montane forests are notable for their rich biotas and high number of regional and local endemic species. Burning, logging, and other conversion leading to intensive agricultural use are the major threats to these forests.

Orinoco

17. Cordillera La Costa montane forests — Venezuela (Vulnerable; Regionally Outstanding; Highest Priority at Regional Scale)
- 13,481 km²
- Sources: **140, 272**; R. Ford Smith, pers. comm.

The moist montane forests of the Cordillera La Costa are found on several isolated mountains and ranges near the Venezuelan coast. These extend from the Paría Peninsula westward to the Sierra de San Luis of the State of Falcón. The biotas of these peaks have been long isolated from one another and larger blocks of moist forest towards the south (e.g., Orinoco Basin, Venezuelan Andes) by drier habitats in the surrounding lowlands. The flora and fauna contain large numbers of regional and local endemics, and many species show disjunct distributions with populations found in the distant Andes or Amazonia. The Paría forests have particularly strong affinities with the biota of Trinidad and Tobago. Urbanization pressures and risk of destructive fires are high-intensity threats to the ecoregion. The forests of the Paría Peninsula are under great threat from expansion of coffee plantations.

18. Orinoco Delta swamp forests — Venezuela, Guyana (Relatively Stable; Bioregionally Outstanding; Moderate Priority at Regional Scale)
- 31,698 km²
- Sources: 111, 137, 140, 141, 176, **272**

The forests of the Orinoco (Amacuro) Delta are important wildlife habitats and are known to support a number of endemic plant and invertebrate species. Oil extraction, water projects, and dam construction represent intensive threats over the next decade. Excessive palmito harvest also represents a threat.

19. Trinidad & Tobago moist forests — Trinidad & Tobago (Vulnerable; Locally Important; Important at National Scale)
- 4,456 km²
- Sources: **10, 11**, 94, 229, 272

No detailed threat information obtained.

20. Guianan Highlands moist forests — Venezuela, Brazil, Guyana (Relatively Intact; Regionally Outstanding; Moderate Priority at Regional Scale)
- 248,018 km²
- Sources: **140**, 182

The Guianan highlands are recognized as an evolutionary center for plant taxa found in both Amazonia and the Guianan lowland forests. The varied geology and topography of the ecoregion have helped create a wide range of plant communities containing many endemic species. Significant endemism is also seen in birds, reptiles and amphibians, invertebrates, and other taxa. The Guianan highlands typically surround many of the tepuis, emergent plateaus that support unique biotas, and share some species with them. Heavy poaching and commercial exploitation of wildlife are becoming serious threats in some areas.

21. Tepuis — Venezuela, Brazil, Guyana, Suriname, Colombia (Relatively Intact; Globally Outstanding; High Priority at Regional Scale)
- 49,157 km²
- Sources: 106, **140, 204, 272**

Tepuis are sandstone plateaus occurring in an east-west belt from Suriname to Peru and Colombia just east of the Andes. The height of the plateaus increases towards the east with many tepuis reaching several thousand meters above the surrounding lowlands. In Peru and Colombia, some formations rise only tens to hundreds of meters above the surrounding lowlands and yet sustain communities characteristic of tepuis. The biological communities of tepuis are notable for their high levels of endemism (even within single plateaus), examples of relict taxa, and for the many unusual adaptations of species to the nutrient-poor, cool, and soggy environments typical of tepuis summits. Because of their isolation, few tepuis have been effected by human activities. Changes in rainfall patterns from lowland deforestation or acidification from distant industrial activity have the potential to degrade sensitive tepuis ecosystems in the future. High-impact tourism at certain sites, burning, and other degradation impacts pose threats within the next five years. Possible expansion of gold mining operations is also a threat.

Amazonia

22. Napo moist forests—Peru, Ecuador, Colombia (Relatively Stable; Globally Outstanding; Highest Priority at Regional Scale)
- 369,847 km²
- Sources: **68, 156,** 221-224, 231, **272**

Biological surveys of different taxa in the Napo provide evidence that this ecoregion contains one of the richest biotas in the world. The entire western arc of the Amazon, particularly the areas near the foothills of the Andes are known for their extraordinary diversity that has been attributed to a number of factors including the high and relatively aseasonal rainfall; the complex topography and soils; vast meandering river systems that create habitat mosaics; and complex biogeographic histories. In the Napo ecoregion, hydrocarbon extraction and associated roadbuilding have caused degradation and fragmentation, and have accelerated these processes by facilitating further colonization. Virtually all of the Ecuadorian portion of the Napo is open for oil leasing. Border controversies between Ecuador and Peru have spurred further colonization in attempts to claim disputed territory.

23. Macarena montane forests—Colombia (Vulnerable; Regionally Outstanding; Highest Priority at Regional Scale)
- 2,366 km²
- Sources: **272**

There was considerable debate as to whether this ecoregion should be incorporated into the surrounding Napo moist forests ecoregion. Although much of the ecoregion is legally protected, progress on a major roadbuilding initiative could increase colonization, legal and illegal commercial activities, and logging.

24. Japura/Negro moist forests—Colombia, Venezuela, Brazil, Peru (Relatively Intact; Globally Outstanding; High Priority at Regional Scale)
- 718,551 km²
- Sources: **68,** 108, 114, **156,** 199, 221-223, **224,** 231, **272**

The Japura/Negro ecoregion contains a great complexity of forest types including terra firme forests, igapó forests, varzea forests, and swamp forests. Some of the world's largest blackwater river ecosystems occur in this ecoregion. Deforestation, agricultural conversion, and colonization pose threats in the next five to ten years. Road construction also represents a threat over the next two decades. Mining may threaten environmental quality in the region.

25. Uatama moist forests—Brazil, Venezuela, Guyana (Relatively Stable; Bioregionally Outstanding; Moderate Priority at Regional Scale)
- 288,128 km²
- Sources: **68,** 221-223, **224,** 231, **272**

Large monospecific tree plantations, mining, associated human settlement and hunting pressures, and selective logging are major threats to this ecoregion. Paving of forest highways would predictably increase settlement pressure and habitat conversion.

26. Amapá moist forests—Brazil, Suriname (Relatively Stable; Bioregionally Outstanding; Moderate Priority at Regional Scale)
- 195,120 km²
- Sources: **68, 103,** 221-223, **224,** 231, **272**

No detailed threat information obtained.

27. Guianan moist forests—Venezuela, Guyana, Suriname, Brazil, French Guiana (Relatively Stable; Bioregionally Outstanding; Moderate Priority at Regional Scale)
- 457,017 km²
- Sources: **68, 103,** 118, 144, 221-224, **272**

This ecoregion is threatened by logging operations which may expand considerably in the next few years. Roadbuilding by international timber companies will likely spur colonization. Gold mining also poses habitat conversion, pollution, and roadbuilding threats.

28. Paramaribo swamp forests—Suriname (Vulnerable; Bioregionally Outstanding; Moderate Priority at Regional Scale)
- 7,760 km²
- Sources: 140, **272**

Agricultural expansion, especially draining of swamp areas and runoff of agrochemicals, is an urgent threat.

29. Ucayali moist forests — Brazil, Peru (Vulnerable; Globally Outstanding; Highest Priority at Regional Scale)
- 173,527 km²
- Sources: **68, 156, 214,** 221-224, 231, **272**

Biogeographic considerations for the Ucayali moist forests are similar to those for the Napo moist forests. No detailed threat information obtained.

30. Western Amazonian swamp forests — Peru, Colombia (Relatively Stable; Globally Outstanding; Highest Priority at Regional Scale)
- 8,315 km²
- Sources: 167, **272**

These swamp forests are ranked as Globally Outstanding because their biotas are strongly associated with the Globally Outstanding ecoregions of western Amazonia.

31. Southwestern Amazonian moist forests — Brazil, Peru, Bolivia (Relatively Stable; Globally Outstanding; Highest Priority at Regional Scale)
- 534,316 km²
- Sources: **68, 156,** 221-223, **224,** 231, **272**

The southern part of this ecoregion in Bolivia should probably be distinguished as a separate ecoregion on biogeographic grounds according to M. Ribera (pers. comm.). Biogeographic considerations for the Southwestern Amazonian moist forests are similar to those for the Napo moist forests. Hydrocarbon extraction and associated roadbuilding have caused degradation and fragmentation and have accelerated these processes by facilitating further colonization. Logging and mining are also threats. An evaluation of the conservation status of the "Beni moist forests" (M. Ribera and E. Forno, pers. comm.) suggests that this portion of the ecoregion is quite threatened; the Relatively Stable status of the ecoregion as a whole is largely due to the more remote northern portion in Brazil and Peru.

32. Juruá moist forests — Brazil (Relatively Intact; Regionally Outstanding; Moderate Priority at Regional Scale)
- 361,055 km²
- Sources: **68,** 221-223, **224,** 231, **272**

Rapid development of petroleum resources in the region of the Rio Juruá poses a major threat. Oil spill hazards, pipeline construction, and settlement of oilfield workers all are causing habitat conversion and degradation. Unregulated hunting and intensive commercial fishing represent immediate threats to vertebrate faunas.

33. Varzea forests — Brazil, Peru, Colombia (Vulnerable; Globally Outstanding; Highest Priority at Regional Scale)
- 193,129 km²
- Sources: **103, 272**

The Varzea forests of the Amazon Basin represent some of the world's most extensive seasonally inundated forests. The seasonal migration of fish and terrestrial animal populations into the flooded forests represent a globally outstanding ecological phenomenon. A number of endemic species, including birds and primates, occur in these forests. Intensive logging and selective exploitation of the kapok tree (*Ceiba pentandra*) are accelerating deforestation. The Varzea already contains extensive industrial timber infrastructure, which will probably spur further logging. Open floodplains are being converted for cattle ranching. The spread of introduced water buffalo is a threat.

34. Purus/Madeira moist forests — Brazil (Relatively Stable; Locally Important; Important at National Scale)
- 561,765 km²
- Sources: **68,** 74, **103 (southern boundary),** 221-223, **224,** 231

Agricultural colonization, land clearing, logging, associated roadbuilding, and hunting represent the most severe threats to this ecoregion.

35. Rondônia/Mato Grosso moist forests — Brazil, Bolivia (Vulnerable; Regionally Outstanding; High Priority at Regional Scale)
- 645,089 km²
- Sources: 3, **103 (eastern and southwestern boundaries),** 221-224, 231, **272 (southwestern boundary)**

This ecoregion supports a wide range of forest types with many transitional formations located southward towards the Cerrado and Beni savannas. Some regions are reported to support highly diverse communities, particularly for butterflies and plants, with many endemic species. During the workshops, there was considerable debate over whether this ecoregion should be categorized as vulnerable or relatively stable. The consequence of these discussions was that this ecoregion is categorized as a Level II priority even though it falls within the Vulnerable/Regionally Outstanding cell in the priority-setting matrix. Deforestation for agriculture and ranching, mining, and roadbuilding all pose major threats over the next two decades. Small-scale logging, wildlife exploitation, introduction of exotic species, and hydroelectric projects are also threats.

Increasing colonization will likely bring further habitat conversion and plant and wildlife exploitation.

36. Beni swamp and gallery forests—Bolivia (Relatively Stable; Locally Important; Important at National Scale)

- 31,329 km²
- Sources: 82, **272**

Expansion of livestock and agricultural land uses pose medium-intensity threats in the next 20 years.

37. Tapajós/Xingu moist forests—Brazil (Vulnerable; Locally Important; Important at National Scale)

- 630,905 km²

Sources: **68**, 74, **103**, 221-223, **224**, 231

Land-clearing for cattle ranching and selective logging of mahogany represent severe threats to this ecoregion.

38. Tocantins moist forests—Brazil (Endangered; Locally Important; Moderate Priority at Regional Scale)

- 279,419 km²
- Sources: **68**, 74, **103 (southern, northern and eastern boundaries)**, 162-164, 221-223, **224**, 231

The Tocantins moist forests face threats from clearing and cattle grazing, which is facilitated by expanded road access. High-intensity selective logging operations are converting habitat and associated operations are leading to forest fires.

Northern Andes

39. Chocó/Darién moist forests—Colombia, Panama, Ecuador (Vulnerable; Globally Outstanding; Highest Priority at Regional Scale)

- 82,079 km²
- Sources: 61, 65, 73, **91**, 97, 98, 102, 105, 110, 132, 149, 150, **151**, 152, **253, 271, 272**

The Chocó/Darién ecoregion is considered to have one of the world's richest lowland biotas, with exceptional richness and endemism in a wide range of taxa including plants, birds, reptiles and amphibians, and butterflies. Unplanned colonization following the completion of roads and massive logging concessions are major threats to the Chocó forests. Since 1960, over 40 percent of the forests have been cleared or heavily degraded and deforestation rates are accelerating (Salaman 1994). This ecoregion faces serious threats in the next five to ten years from national-level development projects, including dams, roads, seaports, pipelines, and military installations. Currently, intensive logging, human settlement,

mining, wildlife exploitation, and coca cultivation all threaten the ecoregion.

40. Eastern Panamanian montane forests—Panama, Colombia (Vulnerable; Bioregionally Outstanding; Moderate Priority at Regional Scale)

- 2,905 km²
- Sources: 112, 113, 124, 190, 262, **271**, 282

The higher peaks of the Serranía de San Blas, Darién, Majé, and Pirre of central and eastern Panama are covered in tropical cloud forest. Both the flora and fauna of these relatively isolated ranges contain numerous endemic species and represent an unusual assemblage of species with South American and Central American affinities. Habitat and environmental quality are being degraded by mining, and numerous wildlife species are being exploited by overhunting for subsistence and trade.

41. Northwestern Andean montane forests—Colombia, Ecuador (Endangered; Globally Outstanding; Highest Priority at Regional Scale)

- 52,937 km²
- Sources: **64**, 150, **272**

The biotas of the submontane and montane forests of the northern Andes are exceptionally rich in species and have a high proportion of regional and local endemics (i.e., species found only in the region or with very limited geographic ranges). The complex topography, climate, geology, and biogeographic history of the northern Andes have helped create a high turnover in species (i.e., beta diversity) over distance and along steep environmental gradients. This region has some of the highest known levels of beta diversity and local endemism for many taxa (e.g., birds—Terborgh and Winter 1983; Hilty and Brown 1986), even to the point where eastern and western slopes of some of the major inter-Andean valleys in Colombia have substantially different biotas (these patterns are particularly pronounced in southwestern Colombia). For these reasons, separate ecoregions were delineated for several ecologically distinct slopes of the northern Andes and all were considered as Globally Outstanding in terms of their associated biodiversity. Many of the moist montane forests of the Northern Andes are under intense threat from conversion for agriculture and pasture, mining operations, and logging.

42. Western Ecuador moist forests—Ecuador, Colombia (Critical; Regionally Outstanding; Highest Priority at Regional Scale)

- 40,218 km²
- Sources: 12a, **64**, 104, 148, 153, 154, 164, **272**

These forests are rich in species with high levels of local and regional endemism. The biota shares strong affinities with that of the adjacent Chocó/Darién moist forests and the Northwestern Andean moist forests, both globally outstanding ecoregions. Intensive logging of non-reserve areas, road construction, and colonization all pose severe threats to this ecoregion.

43. Cauca Valley montane forests—Colombia (Critical; Globally Outstanding; Highest Priority at Regional Scale)
- 32,412 km²
- Sources: **132, 150, 253, 272**

Biodiversity and threat considerations for this ecoregion are covered under the Northwestern Andean montane forests description.

44. Magdalena Valley montane forests—Colombia (Critical; Globally Outstanding; Highest Priority at Regional Scale)
- 49,322 km²
- Sources: 66, **132, 150,** 151, 152, **253**

Biodiversity and threat considerations for the Magdalena Valley montane forests are covered under the Northwestern Andean montane forests description.

45. Magdalena/Urabá moist forests—Colombia (Endangered; Bioregionally Outstanding; High Regional Priority)
- 73,660 km²
- Sources: 73, 150, 253, 272

No detailed threat information obtained.

46. Cordillera Oriental montane forests—Colombia, Venezuela (Vulnerable; Globally Outstanding; Highest Priority at Regional Scale)
- 66,712 km²
- Sources: **132, 150, 272**

Biodiversity and threat considerations for the Cordillera Oriental montane forests are covered under the Northwestern Andean montane forests description.

47. Eastern Cordillera Real montane forests—Ecuador, Colombia, Peru (Vulnerable; Globally Outstanding; Highest Priority at Regional Scale)
- 84,442 km²
- Sources: **64, 132, 150, 272**

Biodiversity and threat considerations for this ecoregion are covered under the Northwestern Andean montane forests description.

48. Santa Marta montane forests—Colombia (Vulnerable; Regionally Outstanding; Highest Regional Priority)

- 4,707 km²
- Sources: 66, **132, 150,** 151, 152, **253, 272**

The Sierra Nevada de Santa Marta rises to nearly 5800 m from the surrounding Caribbean coastal lowlands of Colombia. This isolated massif was present before the Andes rose and accordingly sustains an unusual biota with complex biogeographic affinities (e.g., Amazonian, Mesoamerican, Cho-coan) and high endemism in most taxa. The natural habitats of this range are severely threatened from agricultural expansion, logging, and burning.

49. Venezuelan Andes montane forests—Venezuela, Colombia (Endangered; Globally Outstanding; Highest Priority at Regional Scale)
- 16,638 km²
- Sources: **140, 142,** 143, **272**

Refer to Northwestern Andean montane forests for biodiversity description. This ecoregion is threatened by logging which continues to expand to higher areas. Agricultural colonization represents a low-intensity threat.

50. Catatumbo moist forests—Venezuela, Colombia (Critical; Locally Important; Moderate Priority at Regional Scale)
- 21,813 km²
- Sources: **272; R. Ford Smith, pers. comm.**

The Catatumbo moist forests are at lower elevations (100–300 m) around the southern end of the Maracaibo depression. These forests are noted by Huber and Alarcon (1988) as having few endemic species but interesting affinities with Amazonian floras. Portions of this ecoregion are regularly inundated and sustain swamp forests. Beef and dairy cattle ranching in particular are intense threats. Draining and channelization of wetlands and periodic oil spills are further threats.

Central Andes

51. Peruvian Yungas—Peru (Endangered; Globally Outstanding; Highest Priority at Regional Scale)
- 188,735 km²
- Sources: **156, 214, 237, 272**

The montane forests of the eastern slope of the Andes in Peru support some of the world's richest montane ecosystems. Regional and local endemism in a wide range of taxa is common. Extensive land clearing, agricultural conversion, and logging severely threaten the ecoregion. Roadbuilding and colonization amplify these threats.

52. Bolivian Yungas—Bolivia, Argentina (Endangered; Regionally Outstanding; Highest Regional Priority)

- 72,517 km²
- Sources: 58, 80, 88, 90, 172, **207**, 237, **265, 272**

Ongoing biological surveys suggest that the montane forests of the Bolivian Yungas maintain high levels of endemism, species richness, and beta diversity. The Bolivian Yungas are being deforested for subsistence agriculture and crop production (e.g., coca, coffee, and tea) (Parker 1990).

53. Andean Yungas—Argentina, Bolivia (Vulnerable; Bioregionally Outstanding; Moderate Priority at Regional Scale)
- 55,457 km²
- Sources: **75**, 233, **245, 272**

No detailed threat information obtained.

Eastern South America

54. Brazilian Coastal Atlantic forests—Brazil (Critical; Globally Outstanding; Highest Priority at Regional Scale)
- 233,266 km²
- Sources: **103**

This ecoregion should likely be subdivided into two or more ecoregions to reflect substantial geographic differences in species assemblages. The Atlantic forests are characterized by extraordinarily species-rich biotas and very high levels of both regional and local endemism. Urbanization, industrialization, agricultural expansion, and associated roadbuilding are most severe threats within the next decade. Logging and wildlife exploitation are also threats.

55. Brazilian Interior Atlantic forests—Brazil, Argentina, Paraguay, (Endangered; Regionally Outstanding; Highest Priority at Regional Scale)
- 803,908 km²
- Sources: **103**, 206, 226, **272**

The Interior Atlantic forests display much complexity and geographic variation and probably should be subdivided into several distinct ecoregions in future analyses (e.g., zone intergrading with the Caatinga, Mata Atlântica proper, forests of southern Bahia and Espírito Santo, and the forests of Rio de Janeiro). These forests face threats from agricultural expansion, colonization, logging, and associated road construction within the next five years.

Tropical Dry Broadleaf Forests
Caribbean

56. Cuban dry forests—Cuba (Endangered; Bioregionally Outstanding; High Priority at Regional Scale)

- 61,466 km²
- Sources: 14, 15, 20, 32, **137**, 145

Clearcutting and selective logging, charcoal production, frequent burning, and slash-and-burn agriculture pose threats to the ecoregion.

57. Hispaniolan dry forests—Haiti, Dominican Republic (Endangered; Bioregionally Outstanding; High Priority at Regional Scale)
- 14,610 km²
- Sources: 89, 121, 129, **203**, 205, 207, 208, 241, **256**, 270, 284

The ecoregion faces threats from clearing for development, firewood gathering, and heavy recreational use.

58. Jamaican dry forests—Jamaica (Endangered; Locally Important; Moderate Priority at Regional Scale)
- 2,189 km²
- Sources: 20, **21, 119**, 190

The ecoregion faces threats from clearing for development, firewood gathering, and heavy recreational use.

59. Puerto Rican dry forests—Puerto Rico (Endangered; Locally Important; Moderate Priority at Regional Scale)
- 1,295 km²
- Sources: 20, 76, **92, 177**

Urban expansion, tourism development, livestock grazing, and predation on native birds by exotic mammals all pose threats to the ecoregion.

60. Bahamian dry forests—Bahamas, Turks & Caicos Islands (Vulnerable; Locally Important; Important at National Scale)
- 9,293 km²
- Sources: **20**, 284

The ecoregion faces threats from tourism development, heavy recreational use, and firewood gathering.

61. Cayman Islands dry forests—Cayman Islands (Endangered; Locally Important; Moderate Regional Priority)
- 230 km²
- Sources: **81, 248**

The ecoregion faces threats from tourism development, heavy recreational use, and firewood gathering.

62. Windward Islands dry forests—Windward Islands (Endangered; Locally Important; Moderate Priority at Regional Scale)
- 431 km²
- Sources: 12, 13, **33**, 40-42, **44-52, 54-56, 92, 127, 175, 202, 220**, 269

Agricultural expansion, grazing, firewood gathering, and development are threats to the ecoregion.

63. Leeward Islands dry forests—Leeward Islands (Critical; Locally Important; Moderate Priority at Regional Scale)
- 182 km²
- Sources: 12, 13, **33**, **40-42**, **44-52**, 54-56, **92**, 127, **175**, **202**, **220**, 269

The ecoregion faces threats from clearing for development, seasonal burning, firewood gathering, and heavy recreational use.

Northern Mexico

64. Baja California dry forests—Mexico (Relatively Stable; Locally Important; Important at National Scale)
- 93 km²
- Sources: 19, **99**, 225, 236

Agricultural expansion and grazing threaten the ecoregion.

65. Sinaloan dry forests—Mexico (Vulnerable; Bioregionally Outstanding; Moderate Priority at Regional Scale)
- 119,184 km²
- Sources: 19, 99, **101**, 225, 236

Coffee plantations, firewood gathering, wildlife exploitation, and grazing are threats to the ecoregion.

66. Tamaulipas/Veracruz dry forests—Mexico (Endangered; Locally Important; Moderate Priority at Regional Scale)
- 40,858 km²
- Sources: **99**, 225, 236

Agricultural expansion and grazing threaten the ecoregion.

Central America

67. Jalisco dry forests—Mexico (Endangered; Regionally Outstanding; Highest Priority at Regional Scale)
- 19,973 km²
- Sources: **99**, 225, 236

The dry forests of Southern Mexico (i.e., the Jalisco and Balsas dry forests) are noted for high levels of regional and local endemism in a wide range of taxa. Urbanization and increasing tourism are high-intensity threats to the ecoregion. Road construction, fruit plantations, and ranching also pose threats. Exploitation of wildlife is a high intensity threat as well.

68. Balsas dry forests—Mexico (Endangered; Regionally Outstanding; Highest Priority at Regional Scale)
- 161,098 km²
- Sources: 99, 225, 236

Biodiversity considerations are discussed under Jalisco dry forests. Agricultural expansion, intensive cultivation for export crops and associated pollution threaten the ecoregion.

69. Oaxacan dry forests—Mexico (Endangered; Bioregionally Outstanding; High Priority at Regional Scale)
- 10,566 km²
- Sources: **99**, 180, 225, 236

No clear consensus was reached as to whether this ecoregion should be ranked as Regionally Outstanding or Bioregionally Outstanding. Conversion for cattle ranching, coffee and citrus plantations, and moderate wildlife exploitation threaten the ecoregion.

70. Veracruz dry forests—Mexico (Critical; Locally Important; Moderate Priority at Regional Scale)
- 35,546 km²
- Sources: 99, 225, 236

Agricultural expansion, intensive grazing, firewood gathering, and burning all severely threaten the ecoregion.

71. Yucatán dry forests—Mexico (Endangered; Locally Important; Moderate Priority at Regional Scale)
- 45,554 km²
- Sources: 99, 225, 236

Agricultural expansion, expanding citrus plantations, urbanization, firewood gathering, and grazing pose threats to the ecoregion.

72. Central American Pacific dry forests—El Salvador, Honduras, Nicaragua, Costa Rica, Guatemala (Critical; Bioregionally Outstanding; High Priority at Regional Scale)
- 50,101 km²
- Sources: 60, **99**, 112, 113, 116, 117, 120, **133**, 136, 138, 158, 165, 188, **261**, **273**, 282

The ecoregion is threatened by grazing, burning, agricultural expansion, and exploitative hunting.

73. Panamanian dry forests—Panama (Critical; Locally Important; Moderate Priority at Regional Scale)
- 5,010 km²
- Sources: 112, 117, 200, **271**

The ecoregion is threatened by grazing, burning, and exploitative hunting.

Orinoco

74. Llanos dry forests—Venezuela (Endangered; Locally Important; Moderate Priority at Regional Scale)
- 44,177 km²
- Sources: **140, 142, 272**

No detailed threat information obtained.

75. Trinidad & Tobago dry forests—Trinidad & Tobago (Endangered; Locally Important; Moderate Priority at Regional Scale)
- 251 km²
- Sources: **10, 11, 94, 229, 272**

The ecoregion faces threats from tourism developments, heavy recreational use, and firewood gathering.

Amazonia

76. Bolivian lowland dry forests—Bolivia, Brazil (Critical; Globally Outstanding; Highest Priority at Regional Scale)
- 156,814 km²
- Sources: **171, 191, 272**

Parker et al. (1993) suggest that the dry forests of Bolivia may be among the richest dry forest ecosystems in the world. The biota of the ecoregion has affinities with Amazonia, the Chaco, and the Cerrado as well as containing many endemic species. Agricultural expansion, burning, and grazing pose major threats within the next decade. Increasing wildlife exploitation has the potential to extirpate several target species. Pollution from agricultural and associated human settlements also pose degradation risks to the ecoregion.

Northern Andes

77. Cauca Valley dry forests—Colombia (Critical; Locally Important; Moderate Priority at Regional Scale)
- 5,130 km²
- Sources: **105, 132, 150**, 151, 152, **272**, 283

No detailed threat information obtained.

78. Magdalena Valley dry forests—Colombia (Critical; Locally Important; Moderate Regional Priority)
- 13,837 km²
- Sources: **105, 132, 150**, 151, 152, **272**, 283

No detailed threat information obtained.

79. Patía Valley dry forests—Colombia (Critical; Locally Important; Moderate Priority at Regional Scale)

- 1,291 km²
- Sources: **105, 132, 150**, 151, 152, **272**, 283

No detailed threat information obtained.

80. Sinú Valley dry forests—Colombia (Critical; Locally Important; Moderate Regional Priority)
- 55,473 km²
- Sources: **105, 132, 150**, 151, 152, **272**, 283

No detailed threat information obtained.

81. Ecuadorian dry forests—Ecuador (Critical; Globally Outstanding; Highest Priority at Regional Scale)
- 22,271 km²
- Sources: **64, 147, 270**

The dry forests of the Pacific coast of South America (i.e., Ecuadorian dry forests and Tumbes/Piura dry forests) are known for high levels of both regional and local endemism. Logging and overgrazing present severe degradation threats within the next five years, even within protected areas.

82. Tumbes/Piura dry forests—Ecuador, Peru (Endangered; Globally Outstanding; Highest Priority at Regional Scale)
- 64,588 km²
- Sources: **64, 147, 156, 272**

Biodiversity considerations are discussed under Ecuadorian dry forests. No detailed threat information obtained.

83. Marañón dry forests—Peru (Endangered; Bioregionally Outstanding; High Priority at Regional Scale)
- 14,920 km²
- Sources: **156**, 183, **272; David Neill, pers. comm.**

There are a number of endemic plant and bird species associated with this ecoregion, and other taxa are likely to display similar distribution patterns. Oil palm plantation expansion, cattle ranching, and logging threaten the ecoregion severely in the next five years. Oil exploration and extraction are also major threats.

84. Maracaibo dry forests—Venezuela (Endangered; Locally Important; Moderate Priority at Regional Scale)
- 31,471 km²
- Sources: **272**

No detailed threat information obtained.

85. Lara/Falcón dry forests—Venezuela (Endangered; Locally Important; Moderate Priority at Regional Scale)

- 16,178 km²
- Sources: **140,** 243, 283; **R. Ford Smith, pers. comm.**

No detailed threat information obtained.

Central Andes

86. Bolivian montane dry forests—Bolivia (Critical; Bioregionally Outstanding; High Priority at Regional Scale)
- 39,368 km²
- Sources: 265, 272

Settlement and agricultural conversion have already had dramatic effects on the ecoregion; further expansion seriously threatens remaining fragments of habitat.

CONIFER/TEMPERATE BROADLEAF FORESTS
Temperate Forests
Southern South America

87. Chilean winter-rain forests—Chile (Endan-gered; Regionally Outstanding; Highest Regional Priority)
- 24,937 km²
- Sources: 109, 272; M.K. Arroyo, pers. comm.

These forests are adapted to a Mediterranean climate and share many characteristics with the adjacent Chilean matorral, including its high levels of endemic species. Intensive logging and timber plantations, exotic species invasions, anthropogenic fires, grazing, and firewood gathering are severe threats to the ecoregion.

88. Valdivian temperate forests—Chile, Argentina (Vulnerable; Globally Outstanding; Highest Priority at Regional Scale)
- 166,248 km²
- Sources: **109,** 185, **272**

The Valdivian temperate forests represent one of the world's five major temperate rainforest ecosystems (i.e., Valdivian in Chile, Pacific Northwest of North America, western Black Sea, New Zealand, Tasmania, and some small areas in Japan, Norway, Ireland, and the United Kingdom; Kellogg 1992). The forests of this ecoregion can support extraordinary standing biomass (i.e., stands of very large trees that are characteristic of temperate rainforests) and are known to contain many unusual species and higher taxa. Intensive logging and timber plantations are severe threats to the ecoregion.

89. Subpolar *Nothofagus* forests—Chile, Argentina (Vulnerable; Bioregionally Outstanding; Moderate Priority at Regional Scale)·
- 141,120 km²
- Sources: **109,** 188, 219, **272**

Cold deciduous forests and evergreen swamp forests dominate this ecoregion where still intact. Intensive logging and timber plantations, exotic species invasions (e.g., rabbits removing ground cover), grazing, and firewood gathering are severe threats to the ecoregion.

Tropical and Subtropical Coniferous Forests
Caribbean

90. Cuban pine forests—Cuba (Vulnerable; Regionally Outstanding; Highest Priority at Regional Scale)
- 6,017 km²
- Sources: 14, 15, 20, 32, **131**

The pine forests of Cuba and Hispaniola support a number of endemic plant and animal species (IGACC 1989; Borhidi 1991). Mining, citrus plantations, grazing, and logging severely threaten the ecoregion. Exploitation of threatened parrot populations occurs in western portions of the ecoregion.

91. Hispaniolan pine forests—Haiti, Dominican Republic (Vulnerable; Regionally Outstanding; Highest Priority at Regional Scale)
- 10,833 km²
- Sources: 89, 121, 129, **203,** 205, 207, 232, 241, **256,** 270, 284

Biodiversity considerations are discussed under Cuban pine forests. Degradation and destruction of these forests continues through clearing for agriculture, grazing, firewood collection, and anthropogenic fires.

92. Bahamian pine forests—Bahamas, Turks & Caicos Islands (Vulnerable; Locally Important; Important at National Scale)
- 3,920 km²
- Sources: **20,** 284

The ecoregion faces threats from tourism development, heavy recreational use, and firewood gathering.

Northern Mexico

93. Sierra Juarez pine-oak forests—Mexico, U.S. (Vulnerable; Locally Important; Important at National Scale)
- 24,228 km²
- Sources: **6,** 20, 30, **99,** 112, 113, **174,** 216, 225, 236

Off-road vehicle use, intensive recreational use, settlement pressure, agricultural expansion, and grazing threaten the ecoregion.

94. San Lucan pine-oak forests—Mexico (Relatively Intact; Locally Important; Important at National Scale)

- 895 km²
- Sources: 20, 30, **99**, 112, 113, 216, 225, 236

Agricultural expansion and grazing threaten the ecoregion.

95. Sierra Madre Occidental pine-oak forests—Mexico, U.S. (Endangered; Globally Outstanding; Highest Priority at Regional Scale)
- 204,374 km²
- Sources: **6**, 20, 30, **99**, 112, 113, **173**, 216, 225, 236

The montane forests of the Sierra Madre Occidental represent some of the world's most extensive subtropical coniferous forests. Many plant and animal species are restricted to the diverse forests of this range. Commercial logging, land conversion for cultivation, and overgrazing by livestock pose serious threats to the ecoregion.

96. Central Mexican pine-oak forests (Endangered; Bioregionally Outstanding; High Priority at Regional Scale)
- 3,719 km²
- Sources: 20, 30, **99**, 112, 113, 225, 236

Sheep ranching and overgrazing, agricultural expansion, firewood gathering, and intensive urbanization pose threats. Hunting poses a less severe threat to the integrity of the ecoregion.

97. Sierra Madre Oriental pine-oak forests—Mexico (Relatively Stable; Regionally Outstanding; High Priority at Regional Scale)
- 38,199 km²
- Sources: 20, 30, **99**, 112, 113, 128, 209, 210, 216, 225, 236

The disjunct peaks and ranges of this ecoregion support a number of regional and local endemic species, notably in the birds, conifers, and herpetofauna. Logging, burning, roadbuilding, grazing, and settlement pressures are severe threats to the ecoregion.

98. Veracruz pine-oak forests—Mexico (Critical; Locally Important; Moderate Priority at Regional Scale)
- 5,979 km²
- Sources: 20, 30, **99**, 112, 113, 225, 236

Agricultural expansion, intensive grazing, firewood gathering, and burning all severely threaten the ecoregion.

Central America

99. Mexican transvolcanic pine-oak forests—Mexico (Endangered; Regionally Outstanding; Highest Priority at Regional Scale)
- 72,802 km²
- Sources: 20, 30, **99**, 112, 113, 216, 225, 236

This ecoregion supports rich forests with many regional endemics. Logging, agricultural expansion, firewood gathering, and intensive wildlife exploitation are threats. Sheep ranching and overgrazing, as well as intensive urbanization, pose threats to the ecoregion.

100. Veracruz montane forests—Mexico (Relatively Stable; Bioregionally Outstanding; Moderate Priority at Regional Scale)
- 6,615 km²
- Sources: **99,** 236; R. de la Maza, J. Soberón and G. Castilleja, pers. comm.

No detailed biodiversity information was available for this ecoregion and the appropriateness of its MHT classification is still unclear. Agricultural expansion, intensive grazing, burning, and firewood gathering threaten the ecoregion.

101. Sierra Madre del Sur pine-oak forests—Mexico (Critical; Globally Outstanding; Highest Priority at Regional Scale)
- 41,129 km²
- Sources: 20, 30, **99**, 100, 216, 217, 236

The montane forests of southern Mexico represent some of the world's most diverse and complex subtropical mixed hardwood-conifer forests (WWF and IUCN 1994). The biota of the region is noted for many regional and local endemic species. Roads and associated tourist developments, overgrazing, and exploitive hunting are severe threats to the ecoregion. Urbanization pressures are amplifying these problems and causing further habitat loss.

102. Central American pine-oak forests—Guatemala, El Salvador, Honduras, Mexico, Nicaragua (Vulnerable; Bioregionally Outstanding; Moderate Priority at Regional Scale)
- 127,910 km²
- Sources: 13, 20, 29, 30, **59, 99**, 112, 113, 115, 120, **133**, 134, 136, 138, 155, 159, **216**, 236, 254, 258, 275, 282

The pine-oak forests of the Central American highlands are rapidly disappearing from logging, firewood gathering, wildfires and uncontrolled burning, agricultural expansion, and bark-beetle epidemics that are exacerbated by degradation from logging, grazing, and burning. Road building continues to open areas for exploitation and eventual destruction from expanding human activity.

103. Belizean pine forests—Belize (Relatively Stable; Regionally Outstanding; High Priority at Regional Scale)
- 2,750 km²
- Sources: 20, 59, 112, 113, **123**, 216

Belizean pine forests are dominated by Caribbean pine (*Pinus carribaea*) and require periodic low-intensity burns for their regeneration (Perry 1991). This ecoregion represents one of the few examples of lowland pine forests in the Neotropics. Although not mapped, examples of this ecoregion occur in Mexico and Guatemala. Selective logging and expansion of citrus and banana plantations threaten coastal areas of the ecoregion.

104. Miskito pine forests—Nicaragua, Honduras (Relatively Stable; Regionally Outstanding; High Priority at Regional Scale)
- 15,064 km²
- Sources: 20, 30, **133, 216, 254,** 257

The Miskito pine-savannas represent the largest lowland tropical pine-savannas in the Neotropics. The pine-savannas are dominated by *Pinus carribaea*, a lowland pine dependent on periodic low-intensity burns for its regeneration (Perry 1991). Industrial logging, firewood gathering, wildfires, and uncontrolled burning are threats to the ecoregion.

Eastern South America

105. Brazilian *Araucaria* forests—Brazil, Argentina (Critical; Bioregionally Outstanding; High Priority at Regional Scale)
- 206,459 km²
- Sources: 77, 78, **103, 272 (outside of Brazil); M. Pellerano, N. Vlarty, pers. comm.**

Araucaria are primitive conifers restricted to the Southern Hemisphere. *Araucaria angustifolia* forests typically occur in stands associated with dense tropical moist forests, although pure stands occur. A great diversity of soil types and associated florisitic communities, many without *Araucaria*, occur within the ecoregion. The remaining natural habitats of this ecoregion face severe threats from logging and agricultural expansion within the next five years.

GRASSLANDS/SAVANNAS/SHRUBLANDS
Grasslands, Savannas, and Shrublands
Northern Mexico

106. Central Mexican grasslands—Mexico, U.S. (Unclassified; Bioregionally Outstanding; Unclassified)
- 199,919 km²
- Sources: **4, 6,** 9, 20, 30, **99,** 112, 113, 173, **174,** 225, 236

No detailed threat information obtained.

107. Eastern Mexican grasslands—Mexico (Unclassified; Locally Important; Unclassified)
- 3,674 km²
- Sources: 11, 20, 30, **99,** 112, 113, 209, 210, 225, 236

No detailed threat information obtained. The dark blue area on the ecoregions maps just north of the Eastern Mexican grasslands represent the Laguna Madre wetland complex. Insufficent data were available to incorporate this area into the study.

Central America

108. Tabasco/Veracruz savannas—Mexico (Critical; Locally Important; Moderate Priority at Regional Scale)
- 9,252 km²
- Sources: **99,** 236; **G. Castilleja, pers. comm.**

The origin and maintenance of the Tabasco/Veracruz savannas and the Tehuantepec savannas are considered by some biogeographers to be primarily due to human activities. Ranching, overgrazing, settlement pressures, and associated habitat conversion threaten the few remnants of the ecoregion.

109. Tehuantepec savannas—Mexico (Critical; Locally Important; Moderate Priority at Regional Scale)
- 5,614 km²
- Sources: 20, 30, **99,** 112, 113, 225, 236

Refer to note under Tabasco/Veracruz savannas. Agricultural expansion, intensive grazing, and burning all severely threaten the ecoregion.

Orinoco

110. Llanos—Venezuela, Colombia (Relatively Stable; Bioregionally Outstanding; Moderate Priority at Regional Scale)
- 355,112 km²
- Sources: **140, 142,** 255, **272,** 283

The Llanos represent the largest savanna ecosystem in northern South America. The ecoregion consists of a mosaic of moist gallery forests, dry forests, grasslands, and wetlands. Conversion of forested habitats for agriculture and pasture, draining and channelization of wetlands and aquatic habitats, and frequent burning of habitats during the dry season all threaten the natural habitats of the Llanos.

Amazonia

111. Guianan savannas—Suriname, Guyana, Venezuela, Brazil (Relatively Stable; Bioregionally Outstanding; Moderate Priority at Regional Scale)
- 128,375 km²
- Sources: **103 (Brazil),** 249, 280; **O. Huber, pers. comm.**

Cattle ranching, rice plantations, and other agricultural expansion pose substantial threats within the next decade. Frequent anthropogenic fires increasingly degrade and destroy gallery forests in many areas. Road building and the human activities that follow represent a threat to remote areas.

112. Amazonian savannas — Brazil, Colombia, Venezuela (Relatively Stable; Bioregionally Outstanding; Moderate Priority at Regional Scale)
- 120,124 km²
- Sources: 27, **103,** 130, 144, 189, 218, 249-252, 255, **272**

Amazonian savannas are distributed over a wide geographic area and accordingly display much variation in their biotas and community structures. Savannas with unusual edaphic conditions typically maintain the highest levels of local endemism (Whitmore and Prance 1987). Conservation efforts should emphasize geographic representation of each distinct savanna community. Burning and extraction of white-sand silica are threats to the ecoregion as well as mining, cattle grazing, and frequent burning.

113. Beni savannas — Bolivia (Endangered; Bioregionally Outstanding; High Priority at Regional Scale)
- 165,445 km²
- Sources: 58, **272**

The Beni savannas support a diverse grassland flora and many species of large mammals and birds characteristic of Southern Cone grasslands and savannas. Degradation from development and wildlife exploitation are serious threats to the ecoregion in the next two decades. Overgrazing and frequent burning are continuing severe threats.

Eastern South America

114. Cerrado — Brazil, Bolivia, Paraguay (Vulnerable; Globally Outstanding; Highest Priority at Regional Scale)
- 1,982,249 km²
- Sources: 3, 5, 71, **103,** 160, 161, 186, 187, 249-252, 263, 264, 268, **272 (outside of Brazil),** 274, 277, 281

The Cerrado constitutes one of the largest savanna-forest complexes in the world and contains a diverse mosaic of habitat types and natural communities. We chose to classify the Cerrado as a savanna, rather than a dry forest, due to the mosaic nature of the habitat. Patterns of biodiversity are complex and many regional and local endemic species are present. The scarcity of information on patterns of biodiversity prevented further subdivision of the Cerrado into two or more ecoregions, a biologically justifiable revision. Agricultural expansion, charcoal production, and water

projects pose major threats over the next two decades. Pollution and road construction represent additional threats over the next decade.

115. Chaco savannas — Argentina, Paraguay, Bolivia, Brazil (Vulnerable; Regionally Outstanding; Highest Priority at Regional Scale)
- 611,053 km²
- Sources: 1, 23, 24, 26, 27, 31, 72, **75,** 78, 95, 187, 192, 233, 239, 249-252, **272**

The Chaco supports a diverse flora and fauna with many regional endemics, and a great complexity of habitat types. Excessive grazing by domestic livestock significantly alters community structure and ecological processes and destroys critical aquatic habitats. Wildfires and seasonal burning contribute to degradation and conversion of native habitats. Agricultural expansion threatens some areas.

116. Humid Chaco — Argentina, Paraguay, Uruguay, Brazil (Vulnerable; Locally Important; Important at National Scale)
- 474,340 km²
- Sources: 1, 23, 24, 26, 27, 37, 72, **75, 103,** 233, **272**

No detailed threat information obtained.

117. Córdoba montane savannas — Argentina (Vulnerable; Locally Important; Important at National Scale)
- 55,798 km²
- Sources: **75,** 233

No detailed threat information obtained.

Southern South America

118. Argentine Monte — Argentina (Relatively Stable; Bioregionally Outstanding; Moderate Priority at Regional Scale)
- 197,710 km²
- Sources: **75 (eastern boundary),** 228, 239, **280 (western boundary)**

No detailed threat information obtained.

119. Argentine Espinal — Argentina (Vulnerable; Bioregionally Outstanding; Moderate Priority at Regional Scale)
- 207,054 km²
- Sources: 27, **75,** 228, **246**

No detailed threat information obtained.

120. Pampas — Argentina (Endangered; Bioregionally Outstanding; High Priority at Regional Scale)
- 426,577 km²
- Sources: **75,** 228, 278

Conversion of natural habitats for agriculture and degradation through excessive grazing are important threats. Burning and draining also threaten remaining natural communities.

121. Uruguayan savannas—Uruguay, Brazil, Argentina (Vulnerable; Bioregionally Outstanding; Moderate Priority at Regional Scale)
- 336,846 km²
- Sources: **75**, 86, **103**, **201**, 206, **272**, 278

No clear consensus emerged as to the most appropriate conservation status for the Uruguayan savannas. The degree of habitat degradation from grazing, burning, draining, and exotic species was subject to much debate. Although the ecoregion was eventually categorized as Vulnerable, several authors of this report and other experts felt that an Endangered or Critical categorization was more appropriate. Excessive grazing by livestock and conversion of natural habitats for agriculture represent the primary threats to this ecoregion. Logging reduces forest cover in the western portion of the ecoregion.

Flooded Grasslands
Caribbean

122. Cuban wetlands—Cuba (Endangered; Regionally Outstanding; Highest Priority at Regional Scale)
- 5,345 km²
- Sources: 14-16, 32, 87, **131**, 240

The Zapata Swamp on the southern coast of Cuba is noted for its large size and endemic species. Draining and agricultural expansion, agricultural pollution, charcoal production, grazing, peat extraction, and exotic invasions all pose severe threats to the ecoregion.

123. Enriquillo wetlands—Haiti, Dominican Republic (Vulnerable; Regionally Outstanding; Highest Priority at Regional Scale)
- 574 km²
- Sources: 87, 89, **203**, 205, 207, 240, **256**, 284

Water diversion for irrigation, draining, grazing, firewood gathering, and crocodile poaching are threats in the ecoregion.

Northern Mexico

124. Central Mexican wetlands—Mexico (Critical; Regionally Outstanding; Highest Priority at Regional Scale)
- 362 km²
- Sources: 20, 69, 87, **99**, 240

Central Mexican wetlands are not mapped on any of the ecoregion maps. Conversion, draining, grazing, burning, and pollution threaten these wetlands.

Central America

125. Jalisco palm savannas—Mexico (Critical; Bioregionally Outstanding; High Priority at Regional Scale)
- 555 km²
- Sources: 20, 30, 79, **99**, 112, 113, 225, 236, 240

Palmar formations are dominated by various species of palm and occur in many localities throughout Mexico, with much variation in community structure and composition (Rzedowski 1978). The Jalisco and Veracruz ecoregions represent two of the largest palmar formations. Conservation efforts should strive to preserve each of the many distinct types of palmar communities found in Mexico. Some palmars do not flood (Rzedowski 1978) and would be more appropriately categorized as dry savannas (in the grasslands, savannas, and shrublands MHT). Agricultural expansion, cattle grazing, and burning threaten the ecoregion.

126. Veracruz palm savannas—Mexico (Critical; Bioregionally Outstanding; High Priority at Regional Scale)
- 7,518 km²
- Sources: 20, 30, 79, **99**, 112, 113, 225, 236, 240

Agricultural expansion, cattle grazing, and burning threaten the ecoregion.

127. Quintana Roo wetlands—Mexico (Relatively Intact; Bioregionally Outstanding; Important at National Scale)
- 2,452 km²
- Sources: 20, 69, 79, 87, **99**, 179, 240

No detailed threat information obtained.

Orinoco

128. Orinoco wetlands—Venezuela (Relatively Stable; Bioregionally Outstanding; Moderate Priority at Regional Scale)
- 6,403 km²
- Sources: 87, 102, 113, 137, 140, 141, 176, 240, **272**

Flooded grasslands occur in a habitat mosaic with swamp forests and mangroves in the Orinoco (Amacuro) Delta. Oil extraction, water projects, and dam construction represent intensive threats over the next decade.

Amazonia

129. Western Amazonian flooded grasslands—Peru, Bolivia (Relatively Stable; Bioregionally Outstanding; Moderate Priority at Regional Scale)
- 10,111 km²
- Sources: 197, 218, **272**

No detailed threat information obtained.

130. Eastern Amazonian flooded grasslands—Brazil (Vulnerable; Bioregionally Outstanding; Moderate Priority at Regional Scale)
- 69,533 km²
- Sources: **103**, 197, 218, **272**

Extensive areas of these flooded grasslands are increasingly being converted to pasture, particularly in the Amazon Delta region.

131. São Luis flooded grasslands—Brazil (Endangered; Locally Important; Moderate Priority at Regional Scale)
- 1,681 km²
- Sources: **103**, **272**

Some experts question whether this area deserves recognition as an ecoregion. No detailed threat information obtained.

Northern Andes

132. Guayaquil flooded grasslands—Ecuador (Endangered; Bioregionally Outstanding; High Priority at Regional Scale)
- 3,617 km²
- Sources: 240, **272**

Flooded grasslands are reported to occur east of the Río Daule and west of Guayaquil (UNESCO 1980). No detailed threat information obtained.

Eastern South America

133. Pantanal—Brazil, Bolivia, Paraguay (Vulnerable; Globally Outstanding; Highest Priority at Regional Scale)
- 140,927 km²
- Sources: 8, **103**, 240, **272**

The Pantanal represents one of the world's largest wetland complexes and supports abundant populations of wildlife. The region is noted for huge seasonal aggregations of water birds and caiman. The Pantanal is comprised of a mosaic of flooded grasslands and savannas, gallery forests, and dry forests. During the rainy season over 80 percent of the region floods, a process that helps to modify the severity and frequency of floods downstream along the Río Paraguay. Agricultural expansion, charcoal production, and water projects pose severe threats over the next two decades (Bucher et al. 1993). Pollution and road construction are additional threats projected for the next decade.

134. Paraná flooded savannas—Argentina (Endangered; Bioregionally Outstanding; High Priority at Regional Scale)
- 36,452 km²
- Sources: **75**, 240

Expansion of agriculture and pasture degrade or destroy natural habitats. Channelization and draining alter critical flooding and nutrient cycles that help maintain natural communities.

Montane Grasslands
Central America

135. Mexican alpine tundra—Mexico (Vulnerable; Bioregionally Outstanding; Moderate Priority at Regional Scale)
- 147 km²
- Sources: 20, 30, **99**, 112, 113, 225, 236

Alpine tundra, or zacotonal, occurs near the summits of large volcanoes along the transvolcanic range of central Mexico. Several bird, mammal, invertebrate, and plant species are restricted to this ecoregion. No detailed threat information obtained.

136. Costa Rican paramo—Costa Rica (Relatively Stable; Bioregionally Outstanding; Moderate Priority at Regional Scale)
- 31 km²
- Sources: 62, 139, 165, **261**

Refer to Santa Marta paramo description for biodiversity information. The Costa Rican paramo is the northernmost example of this formation. Extensive and frequent anthropogenic fires threaten portions of the Costa Rican paramo.

Northern Andes

137. Santa Marta paramo—Colombia (Vulnerable; Globally Outstanding; Highest Priority at Regional Scale)
- 1,329 km²
- Sources: 110, **272**, 276

Paramo formations are restricted to high peaks and mountain ranges of the tropics. Although paramo-like formations occur on isolated peaks and ranges of eastern and central Africa and on Mt. Kinabalu in Borneo, this habitat type is most extensive in the Neotropics. Paramo plants and animals display remarkable adaptations to the cold and drying conditions of high elevations. Many paramo species are restricted to these habitats, and local endemism occurs in a wide range of taxa, particularly on more isolated peaks. Paramo formations are threatened by frequent burning, grazing, and conversion for agriculture in some areas. The loss of downslope forests can contribute to the expansion of paramo in some cases (UNESCO 1980). A discussion of threats to paramo formations in the Neotropics can be found in Balslev and Luteyn (1992).

138. Cordillera de Mérida paramo—Venezuela (Relatively Stable; Globally Outstanding; Highest Priority at Regional Scale)
- 3,518 km²
- Sources: **272, 276**

Refer to Santa Marta paramo description for biodiversity information. No detailed threat information obtained.

139. Northern Andean paramo—Colombia, Ecuador (Vulnerable; Globally Outstanding; Highest Priority at Regional Scale)
- 58,806 km²
- Sources: 97, **272, 276**

Refer to Santa Marta paramo description for biodiversity information. No detailed threat information obtained.

Central Andes

140. Cordillera Central paramo—Ecuador, Peru (Vulnerable; Globally Outstanding; Highest Priority at Regional Scale)
- 14,128 km²
- Sources: 28, **156, 272, 276**

Refer to Santa Marta paramo description for biodiversity information. No detailed threat information obtained.

141. Central Andean puna—Bolivia, Argentina, Chile, Peru (Vulnerable; Regionally Outstanding; Highest Priority at Regional Scale)
- 183,868 km²
- Sources: 28, **75** (southeastern boundary), 96, **207, 214,** 242, **272**

Puna formations are montane grasslands of the central and southern High Andes and consist of various communities of bunchgrasses, small shrubs, trees, and herbaceous plants. Intact vertebrate faunas are characterized by camelids (e.g., vicuña, alpaca), condors, and a variety of high altitude rodents, marsupials, canids, and birds. The puna formations of LAC represent one of the world's largest complexes of montane grasslands, the other being in central Tibet. Puna formations have been extensively altered for agriculture and are degraded in many areas through grazing of domestic livestock (e.g., llamas, goats, sheep), burning, and the collection of firewood.

142. Central Andean wet puna—Peru, Bolivia, Chile (Vulnerable; Regionally Outstanding; Highest Priority at Regional Scale)
- 188,911 km²
- Sources: 28, 96, **207, 214, 272**

Refer to Central Andean puna description for biodiversity and threat information.

143. Central Andean dry puna—Argentina, Bolivia, Chile (Vulnerable; Regionally Outstanding; Highest Priority at Regional Scale)
- 232,958 km²
- Sources: 28, 57, 96, **272**

Refer to Central Andean puna description for biodiversity and threat information. The dry puna contains saline and soda lakes that support populations of flamingo and other wildlife. Soda lake ecosystems represent an unusual ecological phenomenon.

Southern South America

144. Southern Andean steppe—Argentina, Chile (Relatively Stable; Locally Important; Important at National Scale)
- 198,643 km²
- Sources: 31, 57, 70, **75** (eastern boundary), 96, **272** (western boundary)

No detailed threat information obtained.

145. Patagonian steppe—Argentina, Chile (Vulnerable; Regionally Outstanding; Highest Priority at Regional Scale)
- 474,757 km²
- Sources: 31, 57, 70, **75** (eastern boundary), 126, 233, **272** (western boundary)

Both the Patagonian steppe and Patagonian grasslands ecoregions are considered under montane grasslands because their ecological dynamics and conservation requirements most closely match those for that MHT. The Patagonian steppe support regionally distinctive communities of mammals and birds, including many unusual higher taxa. Overgrazing and associated erosion, conversion for agriculture, and burning are major threats.

146. Patagonian grasslands—Argentina, Chile (Vulnerable; Bioregionally Outstanding; Moderate Priority at Regional Scale)
- 59,585 km²
- Sources: 31, 57, 70, **75,** 230, 233

The Malvinas (Falkland) Islands are included within this ecoregion, although further biogeographic analysis may prove their inclusion to be inappropriate.[3] See also note under Patagonian steppe. Excessive grazing by livestock and introduced herbivores is a major threat.

---

3. A dispute concerning sovereignty over the islands exists between Argentina, which claims this sovereignty, and the U.K., which administers the islands.

XERIC FORMATIONS
Mediterranean Scrub
Northern Mexico

147. California coastal sage-chaparral — Mexico, U.S. (Critical; Globally Outstanding; Highest Priority at Regional Scale)
- 27,104 km²
- Sources: **4, 6**, 9, 19, 20, 30, **99**, 112, 113, 173, **174**, 209, 210, 225, 234

The coastal sage-chaparral communities of California are categorized as Globally Outstanding because (a) Mediterranean scrub communities are rare in the world, occurring only in five relatively small coastal areas characterized by cool winter rains and warm, dry summers with abundant fog (i.e., the Mediterranean, California coastal chaparral, Chilean matorral, Fynbos of southern Africa, and the heathlands of southwestern Australia); (b) the California coastal chaparrals are the only example of Mediterranean scrub ecosystems in North America; and (c) these ecoregions are extraordinarily rich in species given the relatively low rainfall of these regions, largely because of the high levels of beta diversity and associated local endemism found in these communities. Many plant and invertebrate groups display high diversity in these communities (e.g., the highest known species richness in bees in North America occurs in this ecoregion) and there is a high degree of endemism in many taxa, including vertebrates. The ecoregion is severely threatened by rapidly expanding suburban sprawl in southern California, exotic species, and frequent anthropogenic fires.

Central Andes

148. Chilean matorral — Chile (Endangered; Globally Outstanding; Highest Priority at Regional Scale)
- 141,643 km²
- Sources: **234, 266, 272; M.K. Arroyo, pers. comm.**

The Chilean matorral represents the only Mediterranean scrub ecoregion in all of South America, and it is only one of five such ecosystems in the world (see California coastal sage-chaparral). The biota is characterized by high levels of species richness, regional and local endemism (particularly in plants), and beta diversity in a wide range of taxa. The ecoregion is threatened by conversion for agriculture, pasture, and development, frequent anthropogenic fires, exotic species, and grazing.

Deserts and Xeric Shrublands
Caribbean

149. Cuban cactus scrub — Cuba (Vulnerable; Locally Important; Nationally Important Priority)
- 3,044 km²
- Sources: 14, 15, 32, **131**

Grazing, woodcutting, and the conversion and resource exploitation associated with increased urbanization pose threats to the ecoregion for the foreseeable future.

150. Cayman Islands xeric scrub — Cayman Islands (Endangered; Locally Important; Moderate Regional Priority)
- 32 km²
- Sources: **81, 248**

Habitats are threatened by development for tourism facilities and overgrazing.

151. Windward Islands xeric scrub — Windward Islands (Endangered; Locally Important; Moderate Priority at Regional Scale)
- 944 km²
- Sources: 12, 13, **33, 41-52, 54-56**, 92, 122, 127, 175, 202, 284

Grazing, woodcutting, and the conversion and resource exploitation associated with increased urbanization pose threats to the ecoregion for the foreseeable future.

152. Leeward Islands xeric scrub — Leeward Islands (Critical; Locally Important; Moderate Priority at Regional Scale)
- 1,521 km²
- Sources: 12, 13, **33, 41, 42, 44-52, 54-56**, 92, 122, **127, 175**, 202, **220**, 284

Grazing, woodcutting, and the conversion and resource exploitation associated with increased urbanization pose threats to the ecoregion for the foreseeable future.

Northern Mexico

153. Baja California xeric scrub — Mexico (Relatively Intact; Bioregionally Outstanding; Important at National Scale)
- 72,377 km²
- Sources: 9, 20, 30, **99**, 112, 113, 173, 210, 225, 236

Off-road vehicle use, exploitative hunting, and illegal logging of boojum trees (*Idria columnaris*) threaten the ecoregion.

154. San Lucan mezquital — Mexico (Relatively Intact; Locally Important; Important at National Scale)
- 2,226 km²
- Sources: **99, 236**

No detailed threat information obtained.

155. Western Mexican mezquital — Mexico, U.S. (Unclassified; Locally Important; Unclassified)

- 22,894 km²
- Sources: **4, 6,** 9, 18, 19, 20, 30, **99,** 112, 113, 173, **174,** 225, 236

No detailed threat information obtained.

156. Sonoran xeric scrub—Mexico, U.S. (Relatively Stable; Regionally Outstanding; High Priority at Regional Scale)

- 232,339 km²
- Sources: **4, 6,** 9, 19, 20, 30, **99,** 112, 113, 173, **174,** 209, 216, 225, 236

This ecoregion is categorized as Regionally Outstanding because of its exceptionally rich desert flora, subregional endemism in some taxa (e.g., Cactaceae), and unusual floristic communities (e.g., boojum [*Idria columnaris*] deserts of Baja California). Irrigation, cattle ranching, fuelwood extraction, and hunting pose serious threats to the ecoregion.

157. Northern Sonoran cactus scrub—Mexico, U.S. (Relatively Stable; Globally Outstanding; Highest Priority at Regional Scale)

- 97,962 km²
- Sources: **4, 6,** 9, 19, 20, 30, **99,** 112, 113, 173, **174,** 210, 225, 236

The cactus scrub communities of the northern Sonora desert have some of the most diverse and unusual desert biotas in the world. Forests of giant cacti (*Cereus* spp.) are notable here and are associated with a rich variety of both plant and animal species. Irrigation, exploitation of wildlife and plants by collectors and dealers, and intensive recreational use threaten the ecoregion.

158. Mexican Interior chaparral—Mexico, U.S. (Unclassified; Locally Important; Unclassified)

- 22,252 km²
- Sources: **4, 6,** 20, 30, **99,** 112, 113, **174,** 225, 236

No detailed threat information obtained.

159. Chihuahuan xeric scrub—Mexico, U.S. (Vulnerable; Locally Important; Important at National Scale)

- 399,446 km²
- Sources: **4, 6,** 19, 20, **99,** 112, 113, 153, 173, **174,** 209, 210, 225, 236

Grazing, extraction of salt and gypsum, and exploitation of water resources are threats to the ecoregion.

160. Central Mexican mezquital—Mexico (Endangered; Locally Important; Moderate Priority at Regional Scale)

- 29,347 km²
- Sources: 4, 6, 9, 19, 20, 30, **99,** 112, 113, 173, 225, 236

Agricultural conversion and overgrazing have heavily affected the ecoregion and will continue to threaten remaining fragments.

161. Eastern Mexican matorral—Mexico (Unclassified; Bioregionally Outstanding; Unclassified)

- 26,684 km²
- Sources: **4, 6,** 9, 20, 30, **99,** 112, 113, 173, **174,** 225, 236

No detailed threat information obtained.

162. Eastern Mexican mezquital—Mexico, U.S. (Vulnerable; Locally Important; Important at National Scale)

- 138,696 km²
- Sources: 4, 6, 9, 20, 30, **99,** 112, 113, 173, 225, 236

Cattle ranching and industrial development along the Mexico/U.S. border have already affected the ecoregion heavily and will continue to pose threats.

163. Central Mexican cactus scrub—Mexico (Unclassified; Bioregionally Outstanding; Unclassified)

- 37,860 km²
- Sources: 2, 20, 30, **99,** 112, 113, 225, 236

No detailed threat information obtained.

Central America

164. Pueblan xeric scrub—Mexico (Critical; Bioregionally Outstanding; High Priority at Regional Scale)

- 6,818 km²
- Sources: 2, 20, 30, **99,** 112, 113, 225, 236

Lime and mineral extraction and overgrazing by sheep and goats pose threats to the ecoregion.

165. Guerreran cactus scrub—Mexico (Vulnerable; Bioregionally Outstanding; Moderate Priority at Regional Scale)

- 5,232 km²
- Sources: 2, 20, 30, **99,** 112, 113, 225, 236

Overgrazing and household firewood gathering are threats to the ecoregion.

166. Motagua Valley thornscrub—Guatemala (Critical; Bioregionally Outstanding; High Priority at Regional Scale)

- 2,363 km²
- Sources: **166; G. Hartshorn, pers. comm.**

Similar dry habitats also occur in Honduras and should perhaps be included in this ecoregion. Goat grazing and burning threaten the ecoregion.

Orinoco

167. Aruba/Curaçao/Bonaire cactus scrub — Netherlands Antilles (Vulnerable; Locally Important; Important at National Scale)
- 407 km²
- Sources: 12, **40, 43, 53,** 127, **140,** 142

Grazing and conversion and degradation from development are threats.

168. La Costa xeric shrublands — Venezuela (Endangered; Locally Important; Moderate Priority at Regional Scale)
- 64,379 km²
- Sources: **140-142, 272; R. Ford Smith, pers. comm.**

Grazing, frequent anthropogenic fires, and conversion for agriculture, pasture, and development are important threats.

169. Araya and Paría xeric scrub — Venezuela (Endangered; Bioregionally Outstanding; High Priority at Regional Scale)
- 5,424 km²
- Sources: **140, 172; R. Ford Smith, pers. comm.**

Grazing, frequent anthropogenic fires, and conversion for agriculture, pasture, and development threaten remaining habitat.

Northern Andes

170. Galapagos Islands xeric scrub — Ecuador (Vulnerable; Globally Outstanding; Highest Priority at Regional Scale)
- 9,122 km²
- Sources: **272**

The Galapagos Islands are widely known for the remarkable adaptive radiation shown by their biota, and its associated historical role in the development of the theory of natural selection and evolution. Both the flora and fauna display a high degree of endemism as would be expected in such an isolated tropical archipelago. Numerous communities, ecological interactions, and species adaptations are unique to the islands. Threats include overgrazing by domestic and feral livestock, predation by exotic species, anthropogenic fires, poaching, and overharvesting of marine resources.

171. Guajira/Barranquilla xeric scrub — Colombia, Venezuela (Endangered; Bioregionally Outstanding; High Priority at Regional Scale)
- 32,404 km²
- Sources: **272**

No detailed threat information obtained.

172. Paraguaná xeric scrub — Venezuela (Endangered; Bioregionally Outstanding; High Priority at Regional Scale)
- 15,987 km²
- Sources: **272; R. Ford Smith, pers. comm.**

No detailed threat information obtained.

Central Andes

173. Sechura desert — Peru, Chile (Vulnerable; Bioregionally Outstanding; Moderate Priority at Regional Scale)
- 189,928 km²
- Sources: 83, 242, **272**

Biodiversity and threat considerations are summarized under the Atacama desert ecoregion.

174. Atacama desert — Chile (Vulnerable; Bioregionally Outstanding; Moderate Priority at Regional Scale)
- 103,841 km²
- Sources: 83, 242, **272**

Some of the more diverse biological communities of the Atacama and Sechura desert ecoregions occur in lomas formations supported by winter fogs which form over cool Pacific Ocean currents. Coastal river valleys that bring water from the distant Andes also are important habitats for terrestrial biodiversity. Overgrazing by domestic livestock, alteration of water flow patterns in river valleys, and firewood collection threaten these ecoregions.

Eastern South America

175. Caatinga — Brazil (Vulnerable; Bioregionally Outstanding; Moderate Priority at Regional Scale)
- 752,606 km²
- Sources: 24, **103**

Caatinga consists of several kinds of tropical thorn scrub ranging from tall scrub forests to savannas dominated by cacti. The flora of this region has affinities with xeric systems in northwestern South America. Moist forests (brejos) are found on some small hills over 500 m altitude and contain many Amazonian species (Whitmore and Prance 1987). Agricultural expansion, grazing, hunting, and burning are major threats that will affect much of the few remaining natural habitats in the next decade.

Restingas
Northern Andes

176. Paraguaná restingas — Venezuela, Colombia (Endangered; Bioregionally Outstanding; High Priority at Regional Scale)

• 15,987 km²
• Sources: **103**

The dune formations of northern South America are much less extensive than those of Brazil. Their floras are also less diverse with lower levels of endemism, yet they are categorized here as Regionally Outstanding because they are the only significant example of dune communities in the northern portion of LAC. Some coastal dune formations also occur on the Yucatán Peninsula in Mexico and elsewhere (193-196), but they were considered too small to be classified as separate restinga ecoregions. No detailed threat information obtained.

Eastern South America

177. Northeastern Brazil restingas—Brazil (Critical; Globally Outstanding; Highest Priority at Regional Scale)
• 10,248 km²
• Sources: **103**

Biodiversity considerations are discussed under Brazilian Atlantic Coast restingas. No detailed threat information obtained.

178. Brazilian Atlantic Coast restingas—Brazil (Critical; Globally Outstanding; Highest Priority at Regional Scale)
• 8,740 km²
• Sources: **103**

The Brazilian restingas represent some of the world's most extensive coastal sand formations. Habitats range from low scrub to open forests, many with rich floras and numerous habitat specialists (i.e., restricted to restingas) and local endemics (i.e., restricted to particular restinga localities). No detailed threat information was obtained other than general threats from conversion for development and recreational vehicles.

Literature Cited in Appendix F

1. Acevedo, C., J. Fox, R. Gauto, T. Granizo, S. Keel, J. Pinazzo, L. Spinzi, W. Sosa, and V. Vera. 1990. *Áreas prioritarias para la conservación en la Region Oriental del Paraguay*. Centro de Datos para la Conservación, Asunción.
2. Anderson, E.F., S. Arias Montes, N.P. Taylor, and A. Cattabriga. 1994. *Threatened cacti of Mexico*. Royal Botanic Gardens, Kew, Richmond, U.K.
3. Aparecida de Brito, M., C. Sobrevila, J.C. Dalponte, G.A. Borges, and T. Grant. 1990. Setting conservation priorities in the state of Mato Grosso, Brazil. Centro de dados para Conservação–Mato Grosso and the Fundação Estadual do Meio Ambiente (FEMA). Unpublished document.
4. Bailey, R.G. 1976. *Ecoregions of the U.S.* Map. United States Department of Agriculture Forest Service, Ogden, Utah.
5. ___. 1983. Delineation of ecosystem regions. *Environmental Management* 7: 365-373.
6. ___. 1989. Explanatory supplement to ecoregions map of the continents. *Environmental Conservation* 16: 307-309.
7. Balslev, H., and J.L. Luteyn, editors. 1992. *Páramo, an Andean ecosystem under human influence*. Academic Press, London.
8. Banks, V. 1991. *The Pantanal, Brazil's forgotten wilderness*. Sierra Club Books, San Francisco.
9. Barbour, M., and W.D. Billings, editors. 1988. *North American terrestrial vegetation*. Cambridge University Press, Cambridge, Massachusetts.
10. Beard, J.S. 1944. The natural vegetation of the island of Tobago, British West Indies. *Ecological Monographs* 14: 135-163.
11. ___. 1946. *The natural vegetation of Trinidad*. Oxford Forestry Memoirs, No. 20. Clarendon Press, Oxford.
12. ___. 1949. *The natural vegetation of the Windward and Leeward Islands*. Oxford Forestry Memoirs, No. 21. Clarendon Press, Oxford.
12a. Best, B.J., editor. 1992. *The threatened forests of southwest Ecuador*. Biosphere Publications, Leeds, U.K.
13. Bibby, C.J., N.J. Collar, M.J. Crosby, M.F. Heath, C. Imboden, T.H. Johnson, A.J. Long, A.J. Stattersfield, and S.J. Thirgood. 1992. *Putting biodiversity on the map: Priority areas for global conservation*. International Council for Bird Preservation, Cambridge, U.K.
14. Borhidi, A. 1991. *Phytogeography and vegetation ecology of Cuba*. Akadémia Kiadó, Budapest.
15. ___, and O. Muñiz. 1970. Cuba: map of the natural potential vegetation. In A. Borhidi, *Phytogeography and vegetation ecology of Cuba*. Map at a scale of 1:1,250,000. Akadémia Kiadó, Budapest.
16. ___, O. Muñiz, and E. Del Risco. 1993. Plant communities of Cuba. I. Fresh and salt water, swamp and coastal vegetation. *Acta Botanica Hungarica* 29: 337- 376.
17. Braatz, S.M. 1982. Draft environmental profile on Jamaica. Department of State, Washington, D.C.
18. Brockman. C., editor. 1978. *Memoria del mapa de cobertura y uso de la tierra*. ERTS-GEOBOL, La Paz, Bolivia.
19. Brown, D.E., and C.H. Lowe. 1983. *Biotic communities of the Southwest*. Map at a scale of 1:1,000,000. United States Department of Agriculture, General Technical Report Rm-78. Rocky Mountain Forest and Experimental Station.
20. ___, R. Reichenbacher, and S. Franson. 1993. A classification system and map of the biotic com-

munities of North America. Unpublished document and map. United States Environmental Protection Agency, Las Vegas, Nevada.

21. Brown, F.M., and B. Heineman. 1972. *Jamaica and its butterflies.* E.W. Classey Limited, London.

22. Brown, K.S., Jr. 1987. Biogeography and evolution of neotropical butterflies. Pages 66-99 in Whitmore, T.C., and G.T. Prance, editors, *Biogeography and quarternary history in tropical America.* Oxford Monographs on Biogeography No. 3. Clarendon Press, Oxford.

23. Bucher, E.H. 1980. Ecología de la fauna chaqueña. Una revisión. *Ecosur* 7: 111-159.

24. ___. 1982. Chaco and Caatinga—South American arid savannas, woodlands and thickets. Pages 47-79 in B.J. Huntley and B.H. Walker, editors, *Ecology of tropical savannas.* Springer-Verlag, Berlin.

25. ___, A. Bonetto, T.P. Boyle, P. Canevari, G. Castro, P. Huszar, and T. Stone. 1993. *Hidrovia: an initial environmental examination of the Paraguay-Paraná Waterway.* Wetlands for the Americas, Manomet, Massachusetts and Buenos Aires.

26. Cabrera, A.L. 1971. Fitogeografía de la República Argentina. *Bol. Soc. Arg. Bot.* 14: 1-42.

27. ___. 1976. Regiones fitogeográficas argentinas. *Enciclopedia Argentina de agricultura y jardinería.* 2nd ed., vol. II.

28. ___, and A. Willink. 1967. *Biogeografía de América Latina.* Series de Biología, Monografía No. 13. Programa Regional de Desarrollo Cientifico y Tecnológico, Organization of American States, Washington, D.C.

29. Campanella, P. 1982. *Perfil ambiental del pais Honduras.* JRB Associates, McLean, Virginia.

30. Campbell, J.A., and W.W. Lamar. 1989. *The venomous reptiles of Latin America.* Cornell University Press, Ithaca, New York.

31. Canevari, M., P. Canevari, G.R. Carrizo, G. Harris, J. Rodriguez Mata, and R.J. Straneck. 1991. *Nueva guía de las aves Argentinas.* Tomo 1. Buenos Aires.

32. Capote-López, R.P., R. Berazaín-Iturralde, and A. Lieva-Sánchez. 1988. Cuba. Pages 317-335 in D.G. Campbell and H.D. Hammond, editors, *Floristic inventory of tropical countries.* New York Botanical Garden, New York.

33. Caribbean Conservation Association. 1980a. *Survey of conservation priorities in the Lesser Antilles: Virgin Gorda, preliminary data atlas.* Eastern Caribbean Natural Area Management Program, Caribbean Conservation Association, the University of Michigan, and the United Nations.

34. ___. 1980b. *Survey of conservation priorities in the Lesser Antilles: St. Vincent, preliminary data atlas.* Eastern Caribbean Natural Area Management Program, Caribbean Conservation Association, the University of Michigan, and the United Nations.

35. ___. 1980c. *Survey of conservation priorities in the Lesser Antilles: Grenada, preliminary data atlas.* Eastern Caribbean Natural Area Management Program, Caribbean Conservation Association, the University of Michigan, and the United Nations.

36. ___. 1980d. *Survey of conservation priorities in the Lesser Antilles: Anegada, preliminary data atlas.* Eastern Caribbean Natural Area Management Program, Caribbean Conservation Association, the University of Michigan, and the United Nations.

37. ___. 1980e. *Survey of conservation priorities in the Lesser Antilles: Saint Barthélemy, preliminary data atlas.* Eastern Caribbean Natural Area Management Program, Caribbean Conservation Association, the University of Michigan, and the United Nations.

38. ___. 1980f. *Survey of conservation priorities in the Lesser Antilles: Saint Martin, preliminary data atlas.* Eastern Caribbean Natural Area Management Program, Caribbean Conservation Association, the University of Michigan, and the United Nations.

39. ___. 1980g. *Survey of conservation priorities in the Lesser Antilles: Saba, preliminary data atlas.* Eastern Caribbean Natural Area Management Program, Caribbean Conservation Association, the University of Michigan, and the United Nations.

40. ___. 1980h. *Survey of conservation priorities in the Lesser Antilles: Bonaire, preliminary data atlas.* Eastern Caribbean Natural Area Management Program, Caribbean Conservation Association, the University of Michigan, and the United Nations.

41. ___. 1980i. *Survey of conservation priorities in the Lesser Antilles: St. Vincent, Grenadines, preliminary data atlas.* Eastern Caribbean Natural Area Management Program, Caribbean Conservation Association, the University of Michigan, and the United Nations.

42. ___. 1980j. *Survey of conservation priorities in the Lesser Antilles: Anguilla, preliminary data atlas.* Eastern Caribbean Natural Area Management Program, Caribbean Conservation Association, the University of Michigan, and the United Nations.

43. ___. 1980k. *Survey of conservation priorities in the Lesser Antilles: Curaçao, preliminary data atlas.* Eastern Caribbean Natural Area Management Program, Caribbean Conservation Association, the University of Michigan, and the United Nations.

44. ___. 1980l. *Survey of conservation priorities in the Lesser Antilles: St. Lucia, preliminary data atlas.* Eastern Caribbean Natural Area Management Program, Caribbean Conservation Association, the University of Michigan, and the United Nations.

45. ___. 1980m. *Survey of conservation priorities in the Lesser Antilles: St. Eustasius, preliminary data atlas.*

Eastern Caribbean Natural Area Management Program, Caribbean Conservation Association, the University of Michigan, and the United Nations.

46. ___. 1980n. *Survey of conservation priorities in the Lesser Antilles: Montserrat, preliminary data atlas.* Eastern Caribbean Natural Area Management Program, Caribbean Conservation Association, the University of Michigan, and the United Nations.

47. ___. 1980o. *Survey of conservation priorities in the Lesser Antilles: Tortola, preliminary data atlas.* Eastern Caribbean Natural Area Management Program, Caribbean Conservation Association, the University of Michigan, and the United Nations.

48. ___. 1980p. *Survey of conservation priorities in the Lesser Antilles: St. Kitts, preliminary data atlas.* Eastern Caribbean Natural Area Management Program, Caribbean Conservation Association, the University of Michigan, and the United Nations.

49. ___. 1980q. *Survey of conservation priorities in the Lesser Antilles: Grenada, Grenadines, preliminary data atlas.* Eastern Caribbean Natural Area Management Program, Caribbean Conservation Association, the University of Michigan, and the United Nations.

50. ___. 1980r. *Survey of conservation priorities in the Lesser Antilles: Barbuda, preliminary data atlas.* Eastern Caribbean Natural Area Management Program, Caribbean Conservation Association, the University of Michigan, and the United Nations.

51. ___. 1980s. *Survey of conservation priorities in the Lesser Antilles: Nevis, preliminary data atlas.* Eastern Caribbean Natural Area Management Program, Caribbean Conservation Association, the University of Michigan, and the United Nations.

52. ___. 1980t. *Survey of conservation priorities in the Lesser Antilles: Guadeloupe, preliminary data atlas.* Eastern Caribbean Natural Area Management Program, Caribbean Conservation Association, the University of Michigan, and the United Nations.

53. ___. 1980u. *Survey of conservation priorities in the Lesser Antilles: Aruba, preliminary data atlas.* Eastern Caribbean Natural Area Management Program, Caribbean Conservation Association, the University of Michigan, and the United Nations.

54. ___. 1980v. *Survey of conservation priorities in the Lesser Antilles: Antigua, preliminary data atlas.* Eastern Caribbean Natural Area Management Program, Caribbean Conservation Association, the University of Michigan, and the United Nations.

55. ___. 1980w. *Survey of conservation priorities in the Lesser Antilles: Martinique, preliminary data atlas.* Eastern Caribbean Natural Area Management Program, Caribbean Conservation Association, the University of Michigan, and the United Nations.

56. ___. 1980x. *Survey of conservation priorities in the Lesser Antilles: Barbados, preliminary data atlas.* Eastern Caribbean Natural Area Management Program, Caribbean Conservation Association, the University of Michigan, and the United Nations.

57. Castellanos A., and R.A. Perez-Moreau. 1945. *Los tipos de vegetación de la República Argentina.* Universidad Nacional de Tucumán, Tucumán, Argentina.

58. Centro de Datos para la Conservación de Bolivia. 1992. *Bolivia–bosques húmedos densos.* Map. Centro de Datos para la Conservación de Bolivia, La Paz.

59. Chapin, M. 1992a. The coexistence of indigenous people and the natural environment in Central America. Map supplement to *Research and Exploration*, Spring 1992. National Geographic Society, Washington, D.C.

60. ___. 1992b. Disappearing forests; disappearing peoples. *Cultural Survival Quarterly* 16: 63-66.

61. CIAT/EMBRAPA-CPAC. 1985. *Land systems map: Physiography, climate, vegetation, topography, and soil of the central lowlands of tropical South America.* Map at a scale of 1:5,000,000. Centro Internacional de Agricultura Tropical/EMBRAPA-CPAC, Bogotá.

62. Cleff, A.M., and A. Chaverri. 1992. Phytogeography of the páramo flora of Cordillera de Talamanca, Costa Rica. Pages 45-49 in H. Balslev and J.L. Luteyn, editors, *Páramo: An Andean ecosystem under human influence.* Academic Press, London.

63. Clinebell, R.R., II, O.L. Phillips, A.H. Gentry, N. Stark, and H. Zuuring. 1995. Prediction of neotropical tree and liana species richness from soil and climatic data. *Biodiversity and Conservation* 4: 56-90.

64. CLIRSEN/DINAF. 1990. *Mapa forestal de la República de Ecuador.* Map at a scale of 1:1,000,000. CLIRSEN/DINAF, Quito.

65. Cochrane, T.T., L.G. Sanchez, L.G. de Azevedo, J.A. Porras, and C.L. Garver. 1985. *Land in tropical America.* Vol. 2. Centro Internacional para la Agricultura Tropical (CIAT) and EMBRAPA-CPAC, Cali, Colombia.

66. Colchester, M. 1994. The new sultans: Asian loggers move in on Guyana's forests. *The Ecologist* 24: 45-52.

67. Collins, M., editor. 1990. *The last rain forests: A world conservation atlas.* Oxford University Press, New York.

68. Conservation International. 1990. *Biological priorities for conservation in Amazonia.* Map at a scale of 1:5,000,000, developed during the "Workshop 90" held in Manaus, Brazil. Conservation International, Washington, D.C.

69. ___. 1992. *Humedales de México: Tipos de vegetación, humedales prioritarios y áreas protegidas.* Map at a scale of 1:3,000,000. Conservation International, Washington, D.C.

70. Coupland, R.T. 1992. Overview of South American grasslands. Pages 363-406 in W. J. Junk, editor, *Ecosystems of the world 8: Grasslands and savannas.* Elsevier Scientific Publishing Company, New York.

71. Coutinho, L.M. 1982. Ecological effects of fire in Brazilian Cerrado. Pages 273-291 in B.J. Huntley and B.H. Walker, editors, *Ecology of tropical savannas.* Springer-Verlag, Berlin.

72. Croizat-Chaley, L. 1976. *Biogeografía analítica y sintética de las Américas.* Vol. I y II. Academia de Ciencias Físicas, Matemáticas y Naturales, Caracas.

73. Cuatrecasas, J. 1958. Aspectos de la vegetación natural de Colombia. *Revista Academica Colombiana de Ciencias* 10: 221-264.

74. Daly, D.C., and G.T. Prance. 1988. Brazilian Amazon. Pages 402-424 in D.G. Campbell and H.D. Hammond, editors, *Floristic inventory of tropical countries.* New York Botanical Garden, New York.

75. Daniele, C., and C. Natenzon. 1994. *Mapa de las regiones naturales de la Argentina.* Administración de Parques Nacionales, Buenos Aires.

76. Dansereau, P. 1966. *Studies on the vegetation of Puerto Rico.* University of Puerto Rico, Los Baños, Puerto Rico.

77. ___. 1993. Amérique du Sud: Classes de formation de la végétation. Unpublished map at a scale of 1:7,927,000.

78. Dasmann, R. 1976. Biogeographical provinces: Understanding whole systems. *The Coevolution Quarterly,* Sausalito, California.

79. Davidson, I., and M. Gauthier. 1993. *Wetland conservation in Central America.* Report No. 93-3. North American Wetlands Conservation Council. Ottawa.

80. de Morales, C.B. 1990. *Bolivia: Medio ambiente y ecología aplicada.* Instituto de Ecología, Universidad Mayor de San Andrés, La Paz, Bolivia.

81. Department of Tourism (Cayman Islands). 1989. *Visitors map: Cayman Islands.* Map at a scale of 1:50,000. Cayman Islands Government and Ordnance Survey, Southampton, U.K.

82. DeVries, P.J. 1987. *The butterflies of Costa Rica and their natural history.* Princeton University Press, Princeton, New Jersey.

83. Di Castri, F. 1968. *Biologie de l'Amérique Australe 4. Esquisse écologique du Chili.* CNRS, Paris.

84. Dietrich, U. 1992. Situation und perspektiven von Naturschutz und Wildnutzung in Uruguay. Z. *Jagdwiss* 38: 42-54.

85. Dirzo, R., and A. Miranda. 1990. Contemporary neotropical defaunation and forest structure, function, and diversity—a sequel to John Terborgh. *Conservation Biology* 4: 444-447.

86. Ducke, A., and G.A. Black. 1953. Phytogeographical notes on the Brazilian Amazon. *Anais Academia Brasil-Ciencias* 25: 1-46.

87. Dugan, P. 1993. *Wetlands in danger: A world conservation atlas.* The World Conservation Union (IUCN), Oxford University Press, New York.

88. Earth Resources Observation Systems Programme/NASA. 1978. *Mapa de cobertura y uso actual de la tierra, Bolivia.* Map at a scale of 1:1,000,000. NASA.

89. Ehrlich, M. 1985. *Haiti country environmental profile.* United States Agency for International Development (USAID), Washington, D.C.

90. Ellemberg, H. 1981. Mapa simplificado de las ecoregiones. In C.E. Brockman, editor, *Perfil ambiental de Bolivia.* Instituto Internacional para el Desarrollo y Medio Ambiente, Washington, D.C.

91. Espinal, T.L.S., and E.M. Montenegro. 1963. *Formaciones vegetales de Colombia.* Instituto Geográfico "Agustín Codazzi", Bogotá.

92. Ewel, J.J., and J.L. Whitmore. 1973. *The ecological life zones of Puerto Rico and the Virgin Islands.* Forest Service Research Paper No. ITF-18. United States Department of Agriculture.

93. FAO. 1981. *Mapa integrado de ecosistemas forestales del Perú, Colombia, y Venezuela.* Dyeline map at a scale of 1:5,000,000. Food and Agriculture Organization, Project FAO/GCP/RLA/081/JAP.

94. ffrench, R. 1991. *The birds of Trinidad and Tobago.* Cornell University Press, Ithaca, New York.

95. Fittkau, E.J. 1969. *The fauna of South America: Biogeography and ecology in South America.* Hillary, New York.

96. Fjeldså, J., and N. Krabbe. 1990. *Birds of the High Andes.* Zoological Museum, Copenhagen.

97. Fletcher, A.G., editor. 1977. *Atlas de Colombia.* Instituto Geográfico "Agustín Codazzi", Bogotá.

98. ___, editor. 1983. *Atlas regional Pacífico.* Instituto Geográfico "Agustín Codazzi", Bogotá.

99. Flores M.G., L.J. Jiménez, S.X. Madrigal, T.F. Takaki, X.E. Hernández, and R.J. Rzedowski. 1971. *Mapa de tipos de vegetación de la República Mexicana.* Map at a scale of 1:2,000,000. Secretaría de Recursos Hidráulicos.

100. Flores-Villela, O., and P. Gerez Fernández. 1989. *Patrimonio vivo de México: Un diagnóstico de la diversidad biológica.* Instituto Nacional de Investigación sobre Recursos Bióticos (INIREB), Mexico City.

101. Foer, G., and S. Olsen. 1992. *Central America's coasts: Profiles and an agenda for action.* United

States Agency for International Development Regional Office for Central America and The Coastal Resources Center of The University of Rhode Island.

102. Forero, E. 1988. Colombia. Pages 355-360 in D.G. Campbell and H.D. Hammond, editors, *Floristic inventory of tropical countries.* New York Botanical Garden, New York.

103. Fundação Instituto Brasileiro de Geografia Estatística (IBGE). 1993. *Mapa de vegetação do Brasil.* Map at a scale of 1:5,000,000. IBGE, Rio de Janeiro.

104. Gentry, A.H. 1977. *Endangered plant species and habitats of Ecuador and Amazonian Peru.* Missouri Botanical Garden, St. Louis, Missouri.

105. ___. 1988. Northwest South America (Colombia, Ecuador and Peru). Pages 393-399 in D.G. Campbell and H.D. Hammond, editors, *Floristic inventory of tropical countries.* New York Botanical Garden, New York.

106. George, U. 1989. Venezuela's islands in time. *National Geographic* 175: 526-561.

107. Goldsmith, F.B. 1974. An assessment of Foseberg and Ellenberg methods for classifying vegetation for conservation purposes. *Biological Conservation* 6: 3-6.

108. Diego Gómez, P.L. 1984. *An integrated system of biogeographical classification.* Unpublished map at a scale of 1:5,000,000. The Nature Conservancy, Washington, D.C.

109. ___. 1985a. Base map for southern cone South America ecoregion map. Unpublished map at a scale of 1:5,000,000; Lambert's Azimuthal Equal Area Projection. The Nature Conservancy, Washington, D.C.

110. ___. 1985b. South America ecoregions and climatic features. Unpublished map. The Nature Conservancy, Washington, D.C.

111. ___. 1985c. South America ecoregion and climatic features. Draft supplement: Major vegetation types and biotic units of South America. Unpublished blue line map. The Nature Conservancy, Washington, D.C.

112. ___. 1985d. Mesoamerica vegetation Map 1. Provisional vegetation types map of Mexico and Central America. Unpublished map at a scale of 1:5,000,000. The Nature Conservancy, Washington, D.C.

113. ___. 1985e. Biotic units Mesoamerica. Unpublished map at a scale of 1:5,000.000. The Nature Conservancy, Washington, D.C.

114. ___. 1985f. An integrated system of biogeographical classifications for South America, north of Amazon: Biotic units. Unpublished map at a scale of 1:5,000,000. The Nature Conservancy, Washington, D.C.

115. ___. 1985g. Biotic units Mesoamerica. Unpublished map at a scale of 1:5,000,000. The Nature Conservancy, Washington, D.C.

116. ___. 1986. Tipos de vegetación de Costa Rica. Unpublished maps at a scale of 1:250,000; Lambert Projection. The Nature Conservancy, Washington, D.C.

117. ___, and W. Herrera. 1986. *Vegetación y clima de Costa Rica* (2 vol.). Editorial Universidad Estatal a Distancia, San José, Costa Rica.

118. Granville, J. 1979. *La Guyane–Planche 12–Végétation.* Map at a scale of 1:1,000,000. Centre d'Études de Géographie Tropicale, Office de la Recherche Scientifique et Technique d'Outre-Mer.

119. Grossman, D.H., S. Iremonger, and D.M. Muchoney. 1993. *Jamaica: Map of natural communities and modified vegetation types.* The Nature Conservancy, Washington, D.C.

120. Guevara-Morán, J.A. 1985. *El Salvador: perfil ambiental.* EMTECSA de C.V., San Salvador.

121. Hager, J., and T.A. Zanoni. 1993. La vegetación natural de la República Dominicana: una nueva clasificación. *Moscosoa* 7: 39-81.

122. Hamilton S.H., J.O. Juvik, and F.N. Scatena. 1993. *Tropical montane cloud forests: Proceedings of an international symposium.* East-West Center Program on the Environment, Honolulu, Hawaii.

123. Hampshire, R.J. 1988a. Belize. Pages 288-289 in D.G. Campbell and H.D. Hammond, editors, *Floristic inventory of tropical countries.* New York Botanical Garden, New York.

124. ___. 1988b. Panama. Pages 310-312 in D.G. Campbell and H.D. Hammond, editors, *Floristic inventory of tropical countries.* New York Botanical Garden, New York.

125. ___. 1988c. El Salvador. Pages 296-298 in D.G. Campbell and H.D. Hammond, editors, *Floristic inventory of tropical countries.* New York Botanical Garden, New York.

126. Hannah, L., D. Lohse, C. Hutchinson, J.L. Carr, and A. Lankerani. 1994. A preliminary inventory of human disturbance of world ecosystems. *Ambio* 23: 246-250.

127. Harris, D.R. 1965. *Plants, animals and man in the outer Leeward Islands, West Indies.* University of California Publications in Geography, Berkeley, California.

128. Harshberger, J. 1958. *Phytogeographic survey of North America.* 2nd ed. Hafner Publishing Company, New York.

129. Hartshorn, G. 1981. *The Dominican Republic country environmental profile.* JRB Associates, McLean, Virginia.

130. Healey, K. 1990. *Amazonia provisional base map.* Map at a scale of 1:4,000,000; Lambert's Azimuthal

Equal Area Projection. International Travel Map Productions, Vancouver.

131. Hernández, J.R. 1989. *Atlas de Cuba: mapa de la vegetación original de Cuba.* Map at a scale of 1:2,000,000. Instituto de Geografía de Cuba, Havana.

132. Hilty, S.L., and W. L. Brown. 1986. *A guide to the birds of Colombia.* Princeton University Press, Princeton, New Jersey.

133. Holdridge, L.R. 1962a. *Mapa ecológico de Honduras.* Map at a scale of 1:1,000,000. Organization of American States, Washington, D.C.

134. ___. 1962b. *Mapa ecológico de Nicaragua.* United States Agency for International Development, Managua, Nicaragua.

135. ___. 1967. *Life zone ecology.* Tropical Science Center, San José, Costa Rica.

136. ___. 1975. *Zonas de vida ecológicas de El Salvador.* Proyecto Desarrollo, Forestal y Ordenación de Cuencas Hidrográficas, Food and Agriculture Organization.

137. ___. 1976. *República de Venezuela: mapa ecológico.* Map at a scale of 1:2,000,000. Ministerio de Agricultura y Cría y el Fondo Nacional de Investigaciones Agropecuarias, Caracas.

138. ___. 1977. *Mapa ecológico de America Central.* Map.

139. Horn, S.P. 1986. *Fire and paramo vegetation in the Cordillera de Talamanca, Costa Rica.* Ph.D dissertation, University of California, Berkeley.

140. Huber, O., and C. Alarcon. 1988. *Mapa de vegetación de Venezuela.* Map at a scale of 1:2,000,000. Ministerio del Ambiente y de los Recursos Naturales Renovables, Caracas.

141. ___, and D. Frame. 1988. Venezuela. Pages 363-371 in D.G. Campbell and H.D. Hammond, editors, *Floristic inventory of tropical countries.* New York Botanical Garden, New York.

142. Hueck, K. 1960. *Mapa de la vegetación de la República de Venezuela.* Map at a scale of 1:2000,000. Instituto Forestal Latinoamericano de Investigación y Capacitación.

143. ___. 1966. *Die Wülder Südamerikas. Okologie, Zusammensetzung und wirtschaftliche Bedeutung. Vegetationsmonographien.* Bd. II. Stuttgart.

144. ___. 1972. *Mapa de la vegetación de América del Sur.* Map at a scale of 1:8,000,000. Printed with the help of the Ministro Federal do Cooperação Federal and with the collaboration of the Centro Federal de Cooperación Económica of the Federal Republic of Germany.

145. IGACC. 1989. *Nueva atlas nacional de Cuba.* Instituto de Geografía de la Academia de Ciencias de Cuba, Havana.

146. Iñigo-Elias, E.E., and M.A. Ramos. 1991. The Pssitacine trade in Mexico. Pages 380-392 in J. G. Robinson and K. H. Redford, editors, *Neotropical wildlife use and conservation.* The University of Chicago Press, Chicago.

147. Instituto Ecuatoriano de Reforma Agraria y Colonización. 1987. *Croquis ecológico nacional.* Map at a scale of 1:1,000,000. Instituto Ecuatoriano de Reforma Agraria y Colonización, Quito, Ecuador.

148. Instituto Ecuatoriano Forestal y de Areas Naturales y Vida Silvestre. 1990. *Mapa del sistema de áreas naturales protegidas en el Ecuador.* Map at a scale of 1:1,000,000. Instituto Ecuatoriano Forestal y de Areas Naturales y Vida Silvestre, Quito, Ecuador.

149. Instituto Geográfico "Agustín Codazzi". 1969. *Atlas de Colombia.* Instituto Geográfico "Agustín Codazzi", Bogotá, Colombia.

150. ___. 1976. *Mapa ecológico de la República de Colombia.* Instituto Geográfico "Agustín Codazzi", Bogotá, Colombia.

151. ___. 1985. *Mapa de bosques.* Instituto Geográfico "Agustín Codazzi", Bogotá, Colombia.

152. ___. 1988. *Suelos y bosques de Colombia.* Map at a scale of 1:3,400,000. Instituto Geográfico "Agustín Codazzi", Bogotá, Colombia.

153. Instituto Geográfico Militar. 1977. *Atlas geográfico de la República del Ecuador.* Instituto Geográfico Militar, Quito, Ecuador.

154. ___. 1978. *Ecuador: Mapa ecológico.* Map at a scale of 1:1,000,000. Instituto Geográfico Militar, Quito, Ecuador.

155. Instituto Geográfico Nacional. 1972. *Atlas nacional de Guatemala.* Instituto Geográfico Nacional, Guatemala City.

156. Instituto Geográfico Nacional. 1987. *Ecoregiones del Perú.* Map at a scale of 1:5,000,000. Instituto Geográfico Nacional, Lima, Peru.

157. Instituto Geográfico Nacional "Ingeniero Pablo Arnoldo Guzmán". 1987. *Mapa básico de la República de El Salvador.* Instituto Geográfico Nacional "Ingeniero Pablo Arnoldo Guzmán", San Salvador.

158. Inventario Nacional de Recursos Físicos. 1966. *Nicaragua: Vegetation.* Report AID/RIC GIPR No. 6, U.S. Army Corps of Engineers, Washington, D.C.

159. Instituto Nicaragüense de Recursos Naturales y del Ambiente (IRENA). 1992. *Ordenamiento ambiental del territorio: Plan de acción forestal.* IRENA, Managua, Nicaragua.

160. IUCN. 1973. *A working system for classification of world vegetation.* International Union for Conservation of Nature and Natural Resources, Gland, Switzerland.

161. ___. 1974. *Biotic provinces of the world: further development of a system for defining and classifying natural regions for purposes of conservation.* International Union for Conservation of Nature and Natural Resources, Gland, Switzerland.

162. ___. 1990. *Inventário de areas úmidas do Brasil.* World Conservation Union, Universidade de São Paulo, and the Ford Foundation, Gland, Switzerland.

163. ___. 1992a. *Protected areas of the world.* World Conservation Union, Cambridge, U.K.

164. ___. 1992b. *II taller regional de humedales.* World Conservation Union, Quito, Ecuador.

165. Janzen, D. H. 1983. *Costa Rican natural history.* University of Chicago, Chicago, Illinois.

166. Junio, C.A. 1982. *Mapa de cobertura y uso actual de la tierra: República de Guatemala.* Instituto Geográfico Nacional de Guatemala, Guatemala City.

167. Junk, W.J. 1977. Ecology of swamps on the middle Amazon. Pages 269-294 in A.J.P. Gore, editor, *Ecosystems of the world: 4B. Swamp, bog, fen and moor.* Elsevier Scientific Publishing Company, New York.

168. Kalliola, R., M. Puhakka, and W. Danjoy, editors. 1993. *Amazonia Peruana: vegetación húmeda tropical en el llano subandino.* Includes map at a scale of 1:2,000,000. Proyecto Amazonia, Universidad de Turku, Oficina Nacional de Evaluación de Recursos Naturales, Lima, Peru.

169. Kappelle, M. 1992. Structural and floristic differences between wet Atlantic and moist Pacific montane Myrsine-Quercus forests in Costa Rica. In H. Balslev and J. L. Luteyn, editors, *Páramo, an Andean ecosystem under human influence.* Academic Press, London.

170. Kellogg, E., editor. 1992. *Coastal temperate rain forests: ecological characteristics, status and distribution worldwide.* Occasional Paper Series, No. 1. Ecotrust, Portland, Oregon.

171. Kiernan, M. 1992. *Bolivia program strategy summary.* World Wildlife Fund, Washington, D.C.

172. Killeen, T.J., E. Garcia, and S.G. Beck. 1993. *Vegetación de Bolivia: Guía de arboles de Bolivia.* Herbario Nacional de Bolivia and the Missouri Botanical Garden, St. Louis, Missouri.

173. Küchler, A.W. 1967. *Vegetation mapping.* Ronald Press, New York.

174. ___. 1985. *National Atlas of the United States of America: Potential natural vegetation.* Map at a scale of 1:7,5000,000. Department of the Interior, U.S. Geological Survey, Reston, Virginia.

175. Lacereda, L.D. 1994. *Conservation and sustainable utilization of mangrove forests in Latin America and Africa regions. Part 1: Latin America.* Mangrove Ecosystems Technical Reports. International Society for Mangrove Ecosystems, International Tropical Timber Organization.

176. Lindeman, J.C., and S.A. Mori. 1988. The Guianas. Pages 376-388 in D.G. Campbell and H.D. Hammond, editors, *Floristic inventory of tropical countries.* New York Botanical Garden, New York.

177. Little, E.L., Jr., and F.H. Wadsworth. 1964. *Common trees of Puerto Rico and the Virgin Islands.* Agriculture Handbook No. 249. United States Department of Agriculture, Forest Service, Washington, D.C.

178. Long, A.J., M.J. Crosby, A.J. Stattesfield, and D.C. Wege. 1994. Towards a global map of biodiversity: patterns in the distribution of restricted-range birds. Birdlife International, Cambridge, U.K. Unpublished document.

179. Lopez, A. 1983. *Reserva de la Biosfera Sian Ka'an. Vegetación y uso del Suelo: Península de Yucatán.* Map at a scale of 1:2,000,000. Centro de Investigaciones de Quintana Roo, Mexico.

180. Lorence, D.H., and A.G. Mendoza. 1988. Central America: Oaxaca, Mexico. Pages 254-268 in D.G. Campbell and H.D. Hammond, editors, *Floristic inventory of tropical countries.* New York Botanical Garden, New York.

181. Lugo, A.E. 1992. Preservation of primary forests in the Luquillo Mountains, Puerto Rico. *Conservation Biology* 8: 1122-1131.

182. Maguire, B. 1970. On the flora of the Guyana highland. *Biotropica* 2: 85-100.

183. Malleux-Orjeda, J. 1975. *Mapa forestal del Perú.* Universidad Nacional Agraria La Molina, Lima, Peru.

185. Mann, G. 1960. Regiones biogeográficas de Chile. *Investigaciones Zoologicas Chile* 6: 15-49.

186. Mares, M.A. 1992. Neotropical mammals and the myth of Amazonian biodiversity. *Science* 255: 976-979.

187. Mathews, E. 1993. *Global vegetation database.* NASA Goddard Space Flight Center, Washington, D.C.

188. McQueen, D.R. 1976. The ecology of *Nothofagus* and associated vegetation in South America. *Tuatara* 22: 38-68.

189. Meier, W. et al. 1962. Vegetation distribution: Peru, Colombia, Venezuela, Mexico, Guatemala, El Salvador, Costa Rica. *World Forestry Atlas.*

190. Ministry of Agriculture of Jamaica. 1984. *Landcover/use classification 1983/84.* Map at a scale of 1:250,000. Ministry of Agriculture, Kingston.

191. Morales Kreuzer, I. 1993. *Monitoreo del bosque en el Departamento de Santa Cruz.* Proyecto de protección de los recursos naturales en el Departamento de Santa Cruz, Cordecruz-KFW-Consorcio IP/CES/KWC.

192. Morello, J. 1968. *La vegetación de la República Argentina, No. 10: Las grandes unidades de vegetación y ambiente del Chaco Argentino.* Buenos Aires.

193. Moreno-Casasola, P. 1990. Sand dune studies on the eastern coast of Mexico. *Proceedings Canadian Symposium on Coastal Sand Dunes.*

194. ___. 1993. Dry coastal ecosystems of the Atlantic coasts of Mexico and Central America. In E. van der Maarel, editor, *Ecosystems of the world. 2B: Dry coastal ecosystems of Africa, America, Asia and Oceania.* Elsevier Scientific Publishing Company, New York.

195. ___, and I. Espejel. 1986. Classification and ordination of coastal sand dune vegetation along the Gulf and Caribbean Sea of Mexico. *Vegetatio* 66: 147-182.

196. ___, and S. Castillo. 1992. Dune ecology on the eastern coast of Mexico. Pages 309-321 in U. Seeliger, editor, *Coastal plant communities of Latin America.* Academic Press, London.

197. Murça-Pires, J., and G.T. Prance. 1985. The vegetation types of the Brazilian Amazon. Pages 110-144 in G.T. Prance and T.E. Lovejoy, editors, *Amazonia.* New York Botanical Garden, New York.

198. Nelson, C. 1988. Honduras. Pages 291-293 in D.G. Campbell and H.D. Hammond, editors, *Floristic inventory of tropical countries.* New York Botanical Garden, New York.

199. Nelson G., and D. Rosen. 1979. *Vicariance biogeography: A critique.* Colombia University Press, New York.

200. Nentwig, W. 1993. *Spiders of Panama.* Sandhill Crane Press, Gainesville, Florida.

201. OAS. 1970. *Cuenca de Río de La Plata: Formaciones de vegetación natural.* Map at a scale of 1:3,000,000. Organization of American States, Washington, D.C.

202. ___. 1984a. *Saint Lucia: Life zones.* Map at a scale of 1:5,000,000. Organization of American States, Washington, D.C.

203. ___. 1984b. *Écologie, République d'Haïti.* Map at a scale of 1:500,000. Organization of American States, Washington, D.C.

204. ___. 1988. *Suriname Planatlas.* Organization of American States and the National Planning Office of Suriname, Washington, D.C.

205. ___. 1991. *République d'Haïti: Programme de développement de la zone frontalière dans le bassin de haut artibonite.* Map at a scale of 1:150,000. Organization of American States, Washington, D.C.

206. ___. 1992. *Uruguay: estudio ambiental nacional.* Organization of American States, Washington, D.C.

207. Oficina Nacional de Evaluación de Recursos Naturales. 1976. *Mapa ecológico de Perú.* Map at a scale of 1:1,000,000. Oficina Nacional de Evaluación de Recursos Naturales, Lima, Peru.

208. Olson, K.P., V.J. Rudolph, L.M. James, M.R. Koelling, and J.L. Whitmore. 1984. *A national forest management plan for the Dominican Republic.* CRIES Project, Michigan State University and USAID, Washington, D.C.

209. Omernik, J.M. 1986. *Ecoregions of the U.S.* Map at a scale of 1:7,500,000. United States Environmental Protection Agency.

210. ___. 1987. Ecoregions of the conterminous U.S. *Annals of the Association of American Geographers* 77: 118-125.

211. Parker, T.A., III. 1990. Conservation priorities in Bolivia. Conservation International, Washington, D.C. Unpublished document.

212. ___, A.H. Gentry, R.B. Foster, L.H. Emmons, and J.V. Remsen, Jr. 1993. *The lowland dry forests of Santa Cruz, Bolivia: A global conservation priority.* RAP Working Papers No. 4. Conservation International, Washington, D.C.

213. Pearson, O.P., and A.K. Pearson. 1982. Ecology and biogeography of the southern rainforest of Argentina. Pages 129-142 in M.A. Mares and H.H. Genoways, editors, *The Pymatuning Symposia in Ecology. Vol 6: Mammalian biology in South America.* Special Publication Series, Pymatuning Laboratory of Ecology, University of Pittsburgh, Linesville, Pennsylvania.

214. Peñaherrera, C., editor. 1989. *Atlas del Perú.* Instituto Geográfico Nacional, Lima.

215. Peres, C.A., and J.W. Terborgh. 1995. Amazonian nature reserves: an analysis of the defensibility status of existing conservation units and design criteria for the future. *Conservation Biology* 9: 34-46.

216. Perry, Jr., J.P. 1991. *The pines of Mexico and Central America.* Timber Press, Portland, Oregon.

217. Peterson, A.T., O.A. Flores-Villela, L.S. León-Paniagua, J.E. Llorente-Bousquets, M.A. Luis-Martinez, A.G. Navarro-Signza, M.G. Torres-Chávez, and I. Vargas-Fernández. 1993. Conservation priorities in Mexico: Moving up in the world. *Biodiversity Letters* 1: 33-38.

218. Pires, J.M,. and G.T. Prance. 1985. The vegetation types of the Brazilian Amazon. Pages 109-145 in G.T. Prance and T.E. Lovejoy, editors, *Amazonia.* Pergamon Press, Oxford, U.K.

219. Pisano, E. 1980. The Magellanic tundra complex. Pages 295-329 in A.J.P. Gore, editor, *Ecosystems of the world. 4B: Swamp, bog, fen and moor.* Elsevier Scientific Publishing Company, New York.

220. Portecop, J. 1975. *Carte écologique de la Martinique.* Map at a scale of 1:75,000. Centre Universitaire des Antilles, Martinique.

221. Prance, G.T. 1973. Phytogeographic support for the theory of Pleistocene forest refuges in the Amazon basin, based on evidence from distribution patterns in Caryocaraceae, Chrysobalanaceae, and Lecythidaceae. *Acta Amazonica* 3: 5-28.

222. ___. 1977. The phytogeographic divisions of Amazonia and their influence on the selection of

biological reserves. Pages 195-213 in G.T. Prance and T. Elias, editors, *Extinction is forever*. New York Botanical Garden, New York.

223. ___. 1982. Forest refuges: evidence from woody angiosperms. Pages 157-158 in G.T. Prance, editor, *Biological diversification in the tropics*. Columbia University Press, New York.

224. ___. 1985. The changing forests. Pages 146-163 in G.T. Prance and T.E. Lovejoy, editors, *Amazonia*. New York Botanical Garden, New York.

225. Ramamoorthy, T.P., R. Bye, A. Lot, and J. Fa. 1993. *Biological diversity of Mexico: Origins and distribution*. Oxford University Press, New York.

226. Raymundo, M.M. 1993. Uso da terra, da floresta e praticas florestais no estado do Rio Grande do Sul. Unpublished document.

227. Redford, K.H. 1992. The empty forest. *BioScience* 42: 412-422.

228. ___, and J.F. Eisenberg. 1992. *Mammals of the Neotropics, vol. 2. The southern cone: Chile, Argentina, Uruguay, Paraguay*. The University of Chicago Press, Chicago.

229. Republic of Trinidad and Tobago. 1990. *Trinidad*. Map at a scale of 1:150,000. Lands and Surveys Division, Ministry of Planning and Mobilization, Port-of-Spain, Trinidad.

230. Ringuelet, R.A. 1955. Ubicación zoogeográfica de las Islas Malvinas. *Rev. Mus. La Plata (N.S.) Zool.* 6: 419-464.

231. Rizzini, C.T. 1963. Nota prévia sobre a divisão fitogeografica do Brasil. *Revista Brasil. Geografia* 1:1-64.

232. Robert, G. 1984. *Végétation de la République de Haïti*. Université Scientifique et Médicale de Grenoble, Grenoble, Switzerland.

233. Vila, A.R., and C. Bertonatti, editors. 1993. *Situación ambiental de la Argentina: Recomendaciones y prioridades de acción*. Boletín Técnico No. 14, Fundación Vida Silvestre Argentina, Buenos Aires.

234. Rundel, P.W. 1979. The Matorral zone of Central Chile. Pages 175-197 in R.L. Specht, editor, *Ecosystems of the world: Mediterranean-type shrublands*. Elsevier Scientific Publishing Company, New York.

235. Rylands, A.B. 1990. Priority areas for conservation in Amazonia. *Trends in Ecology and Evolution* 5: 240-241.

236. Rzedowski, J. 1978. *Vegetación de México*. Editorial Limusa, Mexico City.

237. Saavedra, C.J., and C.H. Freese. 1985. *Biological priorities for conservation in the tropical Andes*. World Wildlife Fund, Washington, D.C.

238. Salaman, P.G.W., editor. 1994. *Surveys and conservation of biodiversity in the Chocó, south-west of Colombia*. Study Report No. 61., Birdlife International, Cambridge, U.K.

239. Schofield, C.J., and E.H. Bucher. 1986. Industrial contributions to desertification in South America. *Trends in Ecology and Evolution* 1: 78-80.

240. Scott, D.A., and M. Carbonell. 1986. *A directory of Neotropical wetlands*. International Union for the Conservation of Nature (IUCN) and the International Waterfowl and Wetlands Research Bureau (IWRB), Cambridge, U.K.

241. SEA/DED. 1990. *La diversidad biológica en la República Dominicana*. Secretaría de Estado de Agricultura/Departamento de Vida Silvestre, Servicio Alemán de Cooperación Social-Técnica and the World Wildlife Fund, Santo Domingo, Dominican Republic.

242. Simmonetti, J.A., and G. Montenegro. 1994. Conservation and use of biodiversity of the arid and semiarid zones of Chile. Presented at the International Workshop "Conservación y uso sostenible de la biodiversidad en zonas áridas y semiáridas de América Latina," March 1994, Guadalajara, Mexico. Unpublished document.

243. Smith, R.F., and M.A. Rivero. 1991. *Ecología del estado Lara*. Biollania Special Edition No. 1., Caracas, Venezuela.

244. Solomon, A.M., and H.H. Shugart, editors. 1993. *Vegetation dynamics and global change*. Chapman & Hall, New York.

245. Solomon, J.C. 1988. Bolivia. Pages 457-463 in D.G. Campbell and H.D. Hammond, editors, *Floristic inventory of tropical countries*. New York Botanical Garden, New York.

246. Soriano, A. 1989. Río de La Plata grasslands. Pages 367-406 in H. Lieth and M.J.A. Werger, editors, *Ecosystems of the world. Vol. 8: Natural grasslands*. Elsevier Science Publishers, Amsterdam.

247. Stiles, F.G., and A.F. Skutch. 1989. *A guide to the birds of Costa Rica*. Cornell University Press, Ithaca, New York.

248. Stoddart, D.R., M.A. Brunt, and J.A. Davies, editors. 1988. *The biogeography and ecology of the Cayman Islands*. Includes maps at a scale of 1:25,000. W. Junk Publishers, Dordecht, The Netherlands.

249. Stone, T.A. 1992. South America's vanishing natural vegetation: Satellites reveal the destructive and accelerating changes humans have brought on South America's environment. *Cultural Survival Quarterly* 16: 67-70.

250. ___, and P. Schlesinger. 1992. Using 1 km resolution satellite data to classify the vegetation of South America. Pages 85-93 in H. G. Lund, R. Paivinen, and S. Thammincha, editors, *Remote sensing and permanent plot techniques for world forest monitoring*. Proceedings of the IUFRO S4.0205 Wacharakitti International Workshop, 13-17 January 1992, Pattayya, Thailand. Ksetsart University, Finland.

251. ___, P. Schlesinger, R.A. Houghton, and G.M. Woodwell. 1994a. A map of vegetation of South America based on satellite imagery. *Photogrammetric Engineering and Remote Sensing* 60: 541-552.

252. ___, P. Schlesinger, R.A. Houghton, and G.M. Woodwell. 1994b. *A map of the vegetation of South America.* The Woods Hole Research Center, Massachusetts.

253. Suarez-Navarro, A.E., et al. 1984. *Bosques de Colombia.* Instituto Geográfico "Agustín Codazzi", Bogotá, Colombia.

254. Sutton, S.Y. 1988. Nicaragua. Pages 301-303 in D.G. Campbell and H.D. Hammond, editors, *Floristic inventory of tropical countries.* New York Botanical Garden, New York.

255. Tamayo, F. 1964. *Ensayo de clasificación de sabanas de Venezuela.* Universidad Central de Venezuela, Caracas.

256. Tasaico, H. 1967. *Mapa ecológico de la República Dominicana.* Unidad de Recursos Naturales de la Union Panamericana.

257. Taylor, B.W. 1962. The status and development of the Nicaraguan pine savannas. *Caribbean Forester* 23: 21-26.

258. Tebbs, M. 1988. Guatemala. Pages 283-284 in D.G. Campbell and H.D. Hammond, editors, *Floristic inventory of tropical countries.* New York Botanical Garden, New York.

259. Terborgh, J. 1992. The maintenance of diversity in tropical forests. *Biotropica* 24: 283-292.

259a. ___, and B. Winter. 1982. Evolutionary circumstances of species with small ranges. Pages 587-600 in G. T. Prance, editor, *Biological diversification in the tropics.* Columbia University Press, New York.

260. Torres, H. 1992. Biological diversity in South America: Conservation, management and sustainable use. Unpublished document.

261. Tosi Jr., J.A. 1969. *República de Costa Rica: Mapa ecológico.* Map at a scale of 1:750,000. Tropical Science Center, San José, Costa Rica.

262. ___. 1971. *Investigación y demonstraciones forestales. Panamá: zonas de vida.* Programa de las Naciones Unidas para el Desarrollo y la Organización de las Naciones Unidas para la Agricultura y la Alimentación.

263. ___. 1983a. *Provisional ecological map of the Republic of Brazil.* Tropical Science Center, San José, Costa Rica. Map at a scale of 1:10,000,000.

264. ___. 1983b. *Provisional life map of Brazil at 1:5,000,000.* Tropical Science Center, San José, Costa Rica.

265. ___, O. Unzueta, L. Holdridge, and A. Gonzalez. 1975. *Mapa ecológico de Bolivia.* Ministerio de Asuntos Campesinos y Agropecuarios, La Paz, Bolivia.

266. Trabaud, L. 1977. Man and fire: impacts on Mediterranean vegetation. In *Ecosystems of the world: Mediterranean type shrublands.* Elsevier Scientific Publishing Company, New York.

268. Udvardy, M.D.F. 1975. *A classification of the biogeographical provinces of the world.* Occasional Paper No. 8. International Union for the Conservation of Nature (IUCN), Gland, Switzerland.

269. ___, S. Brand, and T. Oberlander. 1978. *World biogeographical provinces.* 3rd edition. Whole Earth Access, Berkeley, California.

270. Unión Panamericana. 1967. *Reconocimiento y evaluación de los recursos naturales de la República Dominicana.* Unión Panamericana, Washington, D.C.

271. UNDP. 1970. *Mapa ecológico de Panamá.* Map at a scale of 1:5,000,000. United Nations Development Programme, Panama.

272. UNESCO. 1980. *Vegetation map of South America.* Map at a scale of 1:5,000,000 and accompanying report. UNESCO and the Institut de la Carte Internationale de Tapis Végétal, Toulouse, France.

273. US Army Corps of Engineers. 1966. *Inventario nacional de recursos físicos de la República de Nicaragua.* USAID/RIC GIPR No. 6. US Army Corps of Engineers, Washington, D.C.

274. Veloso, H.P. 1966. *Atlas florestal do Brasil.* Ministério de Agricultura, Rio de Janeiro.

275. Villela, O.F., and P. Pérez. 1988. *Conservación en México: Síntesis sobre vertebrados terrestres, vegetación y uso del suelo.* Instituto Nacional de Investigaciones sobre Recursos Bióticos (INIREB), Mexico City.

276. Vuilleumier, F. 1970. *Insular biogeography in continental regions: The Northern Andes of South America.* University of Massachusetts, Boston.

277. Walter, H., and E. Box. 1976. Global classification of natural terrestrial ecosystems. *Vegetatio* 32: 75-81.

278. WCMC. 1992. *Global biodiversity: Status of Earth's living resources.* World Conservation Monitoring Centre and Chapman & Hall, London.

279. Whitmore, T.C., and G.T. Prance, editors. 1987. *Biogeography and quaternary history in tropical America.* Oxford Monographs on Biogeography No. 3. Clarendon Press, Oxford.

280. Wiggins, I.L., and D.M. Porter. 1971. *Flora of the Galapagos Islands.* Stanford University Press, Stanford, California.

281. Wikstrom, J.H., and R.G. Bailey. 1983. Land systems classification—possible or impossible? *Renewable Resources Journal.*

282. Wilson, L.D., and J.R. Meyer. 1982. *The snakes of Honduras.* Milwaukee Museum Publications, Milwaukee, Wisconsin.

283. WMO/UNESCO/CARTOGRAPHIA. *Annual average amounts of precipitation (mm): Northern Latin America.* Map at a scale of 1:5,000,000; Bipolar Oblique Conic Conformal Projection. World Metereological Organization, UNESCO, and CARTOGRAPHIA.

284. Woods, C.A., editor. 1989. *Biogeography of the West Indies: Past, present, future.* Sandhill Crane Press, Inc., Gainesville, Florida.

285. WWF and IUCN. 1994. *Centers of plant diversity: A guide for their strategy and conservation.* 3 volumes. World Wildlife Fund and World Conservation Union (IUCN) Publication Unit, Cambridge, U.K.

Appendix G

Sources for Remaining Natural Habitat and Protected Area Assessments

Remaining Habitat Information and Databases

During the BSP and WWF LAC Workshops, regional experts had the opportunity to consult a variety of data sources to assess the conservation status of ecoregions. The source, year, geographic coverage, and quality (i.e., level of resolution and accuracy) of the remaining habitat databases obtained varied considerably. For many regions, reliable or recent (less than 10 years old) data either does not exist or is unavailable because the data is being processed as part of an ongoing study. Recent spatial data on remaining natural habitats for non-forest major habitat types (MHTs) are particularly scarce, due, in part, to difficulties in accurately interpreting habitat information based on remote sensing for these MHTs. The lack of standardized interpretations for different degrees of habitat degradation slows the development of useful spatial databases for all MHTs, including forests.

Below we summarize the spatial databases (i.e., hardcopy maps and digital databases) for remaining natural habitats of LAC that were obtained during the course of this study and made available for review by regional experts at the workshops. We also identify forthcoming spatial databases that we believe will assist future conservation analyses. Although our list is not comprehensive for existing LAC spatial databases or imagery, we hope that it can provide a useful guide for future work. A directory for spatial vegetation databases is being developed by the World Resources Institute.

Hardcopy maps or outputs of the following spatial databases were made available to regional experts at the workshops:

BIOMA. 1988. *Mapa de vegetación de Venezuela.* BIOMA, Caracas. Map at a scale of 1:2,000,000.

Eleven major vegetation categories are used, including urban, industrial, and "áreas intervenidas." Digital database obtained from WCMC.

Caribbean Conservation Association. 1980. *Survey of conservation priorities in the Lesser Antilles. Preliminary data atlases.* Eastern Caribbean Natural Area Management Program, Caribbean Conservation Association, the University of Michigan and the United Nations. Twenty-four bulletins with maps on remaining vegetation were available for different islands: Anegada; Tortola; Virgin Gorda; Anguilla; Saint-Martin; Saint-Barthélemy; Saba; St. Kitts; Nevis; St. Eustasius; Antigua; Barbuda; Montserrat; Guadeloupe; Martinique; St. Vincent; Barbados; St. Vincent Grenadines; Grenada Grenadines; Grenada; Aruba; Curaçao; Bonaire. The sources and dates of the remaining habitat data are variable, but are generally not based on remote sensing technology.

Centro de Datos para la Conservación de Bolivia. 1992. *Bolivia – bosques humedos densos.* Centro de Datos para la Conservación de Bolivia. Map.

Chapin, M. 1992. The coexistence of indigenous people and the natural environment in Central America. Map supplement to *Research and Exploration*, Spring 1992. National Geographic Society, Washington, D.C. Map of remaining forest habitat in Central America (approximately for the period of 1987-1992) based on a wide variety of sources including satellite imagery, aerial photographs, forestry maps, and expert analysis.

CLIRSEN/DINAF. 1990. *Mapa forestal de la República de Ecuador.* Map at a scale of 1:1,000,000; Transverse Mercator Projection. Quito, Ecuador. This map depicts 12 classes of vegetation and "áreas antrópicas."

CODEF. 1987. *Mapa de bosques nativos.* Proyecto 3181: Chile, evaluación de la destrucción del bosque nativo, Comite Nacional para Defensa de la Fauna y Flora. Santiago. Map series at a scale of 1:1,000,000. Map series of remaining forest in portions of the Validivian temperate forests, based on aerial photographs 1978-83; Landsat TM satellite imagery 1985-86; and forest plantation maps from the Forestry Institute.

Collins, M., editor. 1990. *The last rain forests: a world conservation atlas*. Oxford University Press, New York. Maps of remaining rainforests of the world based on a wide variety of sources (primarily non remote-sensing sources).

Conservation International. 1994. Prioridades para conservação da biodiversidade de Mata Atlântica do Nordeste. Conservation International, Washington, D.C. Unpublished map at a scale of 1:250,000; Lambert Conformal Conic Projection. Maps of remaining natural habitat for northeast Brazilian Atlantic forests. Based on data from IBGE, Sociedad Nordestina de Ecologia (SNE), Biodiversitas, and SOS Mata Atlântica. Developed as part of a conservation priority-setting exercise for the northeastern Brazilian Atlantic forests.

Department of Tourism, Cayman Islands. 1989. *Visitors Map: Cayman Islands*. Cayman Islands Government and Ordnance Survey, Southampton, England. Hardcopy map at a scale of 1:50,000 that overlays cartographic data layers of development on aerial photographs (1:25,000 for 1978 and 1989) of Cayman Islands.

Fundação Instituto Brasilero de Geografia Estatística (IBGE). 1993. *Mapa de vegetação do Brasil*. Fundação Instituto Brasilero de Geografia Estatística, Rio de Janeiro, Brasil. Map at a scale of 1:5,000,000. General regions of agricultural land or secondary vegetation are also identified on this map. Vegetation types are derived from analyses by Instituto Brasileiro de Meio Ambiente dos Recursos Naturais Renovaveis (IBAMA), which were based, in part, on radar surveys of Amazonia (Projecto RADAM and RADAM-BRASIL. 1973-1983. *Levantamento de recursos naturais 1-32*. Ministério de Minas e Energia, Rio de Janeiro). A similar radar survey of the Colombian Amazon complements the RADAM-BRASIL study (Proyecto Radargrametrico del Amazonas. 1979. *La Amazonia Colombiana y sus recursos*. Bogotá, Colombia). Whitmore and Prance (1987) review results of the radar surveys.

Government of Trinidad and Tobago. 1977. *Map of Trinidad showing forests and related areas*. Digital database, at a scale of 1:150,000, obtained from WCMC.

Grossman, D.H., S. Iremonger, and D.M. Muchoney. 1993. *Jamaica: map of natural communities and modified vegetation types. Jamaica: a rapid ecological assessment*. The Nature Conservancy, Washington, D.C. This study classified and mapped the natural and modified communities of Jamaica using Landsat TM satellite imagery (1988 and 1989), aerial surveys, ground-truthing, available technical literature, and consultations with regional experts and organizations.

Hannah, L., D. Lohse, C. Hutchinson, J.L. Carr, and A. Lankerani, 1994. A preliminary inventory of human disturbance of world ecosystems. *Ambio* 23: 246-250. This study represents a global assessment of human disturbance of natural ecosystems using three classifications: undisturbed; partially disturbed; and human dominated. A wide variety of source materials were used for the analyses including remote sensing analysis, agricultural atlases, IUCN tropical forest cover maps, and maps from the technical literature. The final map grid is 1000 km².

Huber, O., and C. Alarcon. 1988. *Mapa de vegetación de Venezuela*. Ministerio del Ambiente y de los Recursos Naturales Renovables, Caracas, Vene-zuela. Map at a scale of 1:2,000,000. This vegetation map identifies regions with altered or degraded vegetation and is based on a variety of sources and extensive personal experience of the authors.

IGACC. 1989. *Nueva atlas nacional de Cuba*. Instituto de Geografía de la Academia de Ciencias de Cuba. Havana. This detailed atlas contains a map of existing vegetation types and identifies natural vegetation types, secondary vegetation, areas of pasture and agriculture, plantation forests, and patterns of endemism.

Instituto Geográfico "Agustín Codazzi". 1985. *Mapa de bosques*. Instituto Geográfico "Agustín Codazzi", Bogotá, Colombia. This map identifies remaining forest and their forest type, areas of permanent and shifting agriculture, and areas of intervention.

Junio, C.A. 1982. *Mapa de cobertura y uso actual de la tierra, República de Guatemala*. Instituto Geográfico Nacional Guatemala, Guatemala City. Sources of data for this map series include Landsat TM satellite imagery (Bands 4-5-7; 1:250,000; 1975-1978), aerial photographs, and 1960s land-use maps for the Pacific coast developed by IGN.

OAS. 1988. *Suriname Planatlas*. Organization of American States and the National Planning Office of Suriname, Washington, D.C. This atlas includes a map of existing vegetation that identifies ten vegetation types and cultivated and abandoned land. This map is based on a study by the Foundation for the Conservation of Nature in Suriname and the Central Bureau for Aerial Mapping (map at a scale of 1:1,000,000).

SEA/DED. 1990. *La diversidad biológica en la República Dominicana*. Secretaria de Estado de Agricultura/Departamento do Vida Silvestre, Servicio

Alemán de Cooperación Social-Técnica y Fondo Mundial para la Naturaleza (WWF-US). Santo Domingo, Dominican Republic.

SOS Mata Atlântica-INPE. 1992. *Evolução dos remanescentes de florestais e ecosistemas associados do Domínio da Mata Atlântica. 1985-1990.* Fundação SOS Mata Atlântica and INPE, São Paulo, Brazil. Map series: Paraná, 1:1,700,000; Santa Catarina, 1:1,400,000; Espírito Santo, 1:1,000,000; southern Bahia, 1:750,000. Lambert Conformal Conic Projection. Digital database and hardcopy maps of remaining habitat for the southern portion of the Brazilian Atlantic forest, primarily near the coast. Based on Landsat TM satellite imagery — color composites based on Bands 3D, 4G, and 5; 1:250,000.

Stone, T.A., P. Schlesinger, R.A. Houghton, and G.M. Woodwell. 1994. A map of vegetation of South America based on satellite imagery. *Photogrammetric Engineering and Remote Sensing* 60:541-552. Digital database and hardcopy maps of South American vegetation based on NOAA AVHRR Local Area Coverage satellite data and supplemented by higher resolution satellite imagery, photographs, maps, and personal experience for certain regions. These data have a resolution of 1 km at nadir. The majority of images are from 1988 but range from 1987-1991. The level of resolution is coarser for certain regions including the northern Andes, northeastern Brazil (i.e., the Caatinga ecoregion), and the Southern Cone (15- and 25- km resolution data from Global Vegetation Index data from NOAA was used in some places). Data were classified into 13, 39, and 75 nested land cover classes.

UNESCO. 1980. *Vegetation map of South America.* UNESCO and the Institut de la Carte Internationale de Tapis Végétal, Toulouse, France. Map at a scale of 1:5,000,000. Digital and hardcopy database of vegetation types for South America based, in part, on satellite imagery from MSS Landsat 1 and 2 obtained between 1973 and 1976 (scale approximately 1:1,000,000), and on a variety of maps and bioclimatic, pedological, and phytogeographical data. Altered and degraded formations are identified for many vegetation types.

Remaining Natural Terrestrial Habitat Databases (In Progress)

The following brief summary of ongoing mapping projects is intended to help guide future studies towards useful databases. Several detailed mapping projects are currently underway and will likely provide invaluable information for future conservation analyses. The list of studies below is not comprehensive. Some important projects include:

Belize Land Department Land-use Project: The Land Department of the Ministry of Natural Resources of Belize is currently developing detailed land-use classifications and maps for Belize based on a wide variety of data sources including remote sensing.

Birdlife International: Birdlife International (Cambridge, UK) is in the process of publishing *Important Bird Areas of the Neotropics* that identifies critical localities for the conservation of birds in the Neotropics. Remaining habitat and threat information are provided for different areas in addition to biodiversity data.

GEF Cerrado Workshop: A conservation priority-setting workshop for the Cerrado is being sponsored under a World Bank implemented Global Environment Facility (GEF). project. Mapping of remaining habitat for the upcoming workshop is currently being carried out by Conservation International, Funatura, and other groups.

Pathfinder: The Pathfinder survey is sponsored by NASA and the US Environmental Protection Agency (EPA). The project will measure tropical forest cover and change worldwide at three points in time in the early 1970s, 1980s, and 1990s, and will be based largely on Landsat TM, AVHRR, SPOT satellite imagery and data from the technical literature. For the LAC region, the University of New Hampshire (Brazilian Amazonia) and the University of Maryland (non-Brazilian Amazonia) are conducting the analyses, while the EPA (Las Vegas, Nevada) is mapping Mexico, Central America, and the Caribbean. Six classifications are being used: forest; non-forest vegetation; deforested areas; regenerated forest cover; cloud and water cover; and "unknown." Further information on the Central America and Mexico forest distribution maps can be found in the following citation: McIntosh, S.E., K.B. Lannom, and Z. Zhu. 1994. Development of forest distribution maps of Central America and Mexico from AVHRR data: study plan summary. The Southern Forest Experimental Station, Forest Inventory and Analysis Unit (SOFIA), Starkville, Mississippi.

PANAMAZONIA: This study represents a multinational effort on the part of eight Amazonian Basin countries (led by Brazil's National Institute of Space Research — INPE) to map forest cover of the Amazon Basin during the mid- and late 1980s using satellite data. Both IBAMA and INPE have conducted a number of other mapping projects for Brazil and the Amazon that employ satellite data.

Inventories of Forest Resources in Mexico: The Instituto de Geografía (UNAM; SELPER–México) has recently completed an inventory of forest resources of Mexico. Remaining forest habitats have been classified into 39 forest classes using remote sensing technology, available maps, expert analysis, and ground-truthing exercises. Digital databases and 242 maps (1:250,000) are currently available from UNAM.

TREES: The TREES (Tropical Ecosystem Environment observation by Satellite) project (European Space Agency and the Commission of European Communities) is mapping tropical forest cover for the globe based largely on AVHRR satellite data, with nested SPOT and Landsat TM imagery and ground-truthing. Deforestation dynamics will be monitored in later project phases.

WWF and IUCN Centers of Plant Diversity Project: This project identifies localities, protected areas, and regions that are particularly important for the conservation of global plant biodiversity. Remaining habitat and threat information are provided in the descriptions. The volume for the LAC region will soon be published (WWF and IUCN. 1994. Centres of plant diversity: A guide for their strategy and conservation. 3 volumes. IUCN Publication Unit, Cambridge, U.K.).

World Conservation Monitoring Centre (WCMC): WCMC is in the final stages of producing a conservation atlas of tropical forests for the LAC region with maps of remaining forest cover based on a wide variety of data sources including regional vegetation studies, national atlases, and remote sensing data.

Additional information on monitoring and mapping natural habitats using remote sensing in LAC can be found in the following references (this is not a comprehensive list of relevant references):

Canada Centre for Remote Sensing. 1994. *Canada's Tropical Forestry Initiative: an airborne remote sensing project in Latin America.* Energy, Mines and Resources Canada. Descriptive bulletin.

Moran, E.F., E. Brondizio, P. Mausel, and Y. Wu. 1994. Integrating Amazonian vegetation, land-use, and satellite data. *BioScience* 44: 329-338.

Rudant, J.-P. 1994. French Guiana through the clouds: first complete satellite coverage. *Earth Observation Quarterly* 44: 1-6.

Skole, D.L. and C. Tucker. 1993. Tropical deforestation and habitat fragmentation in the Amazon: satellite data from 1978 to 1988. *Science* 260: 1905-1910.

Skole, D.L., W.H. Chomentowski, W.A. Salas, and A.D. Nobre. 1994. Physical and human dimensions of deforestation in Amazonia. *BioScience* 44: 314-322.

Protected Area Databases

Several sources of information on protected areas were provided to regional experts at the workshop. The source, creation date, geographic coverage, and quality (i.e., level of resolution and accuracy) of the protected areas maps and databases provided varied considerably. For many regions, boundaries of protected areas, classifications of their protected status, and general location differed considerably among maps and sources. Only point data was available for some regions. The following maps and reference materials were available to regional experts at the workshops:

Acevedo, C., J. Fox, R. Gauto, T. Granizo, S. Keel, J. Pinazzo, L. Spinzi, W. Sosa, V. Vera. 1990. *Áreas prioritarias para la conservación en la region Oriental del Paraguay.* Centro de Datos Para la Conservación, Asunción, Paraguay.

Aparecida de Brito, M., C. Sobrevila, J.C. Dalponte, G.A. Borges, and T. Grant. 1990. Setting conservation priorities in the state of Mato Grosso, Brazil. Centro de dados para Conservação–Mato Grosso and the Fundação Estadual do Meio Ambiente (FEMA). Unpublished document.

Burkart, R., C. Daniele, C. Natenzon, and F. Ardura. 1994. *El sistema nacional de áreas naturales protegidas de la Argentina: diagnostico de su patrimonio natural y su desarrollo institucional.* Administración de Parques Nacionales, Buenos Aires.

____, and L. Ruiz. 1994. *República Argentina: áreas protegidas de la Argentina.* Administración de Parques Nacionales, Buenos Aires. Dyeline map (no scale or projection given).

Caribbean Conservation Association. 1980. *Survey of conservation priorities in the Lesser Antilles. Preliminary data atlases.* Eastern Caribbean Natural Area Management Program, Caribbean Conser-vation Association, the University of Michigan and the United Nations. Maps of protected areas are included.

Centro de Dados para a Conservação/FEMA. 1990. Proposta de areas potencias para a conservação (Mato Grosso, Brazil). Centro de Dados para a Conservação/FEMA. Unpublished document.

Centro de Datos para la Conservación. 1994. Conservación de la diversidad biológica en Bolivia. Centro de Datos para la Conservación, La Paz, Bolivia. Draft document.

CIAT. 1994. *Legally protected areas.* Map and digital database at a scale of 1:500,000. Centro Internacional de Agricultura Tropical, Cali, Colombia.

CLIRSEN/DINAF. 1990. *Mapa forestal de la República de Ecuador.* Quito, Ecuador. Map at a scale of

1:1,000,000; Transverse Mercator Projection. Protected areas are shown on this map.

Conservation International. 1992. *Humedales de Mexico: Tipos de vegetación, humedales prioritarios y áreas protegidas.* Map at a scale of 1:3,000,000. Conservation International, Washington, D.C.

Departamento de Áreas Protegidas y Centro de Datos para la Conservación (DPNVS). 1993. *República del Paraguay: Plan maestro del sistema nacional de áreas silvestres protegidas del Paraguay – áreas silvestres protegidas.* Map at a scale of 1:2,000,000. Ministerio de Agricultura y Ganadería, Dirección de Parques Nacionales y Vida Silvestre, Asunción, Paraguay.

Fundação Instituto Brasilero de Geografia Estatística-IBGE. 1994. *Unidades de conservação federais do Brasil.* Map at a scale of 1:5,000,000; Polyconic Projection. Fundação Instituto Brasilero de Geografia Estatística-IBGE, Rio de Janeiro.

Fundación Natura. 1992. *Parques nacionales y otras áreas naturales protegidas del Ecuador.* Fundación Natura, Quito.

Gómez Pompa, A. 1994. *Indice de areas protegidas establecidas prioritarias: Reporte preliminar, base de datos: areas protegidas (Mexico).* Preliminary report presented to the World Bank.

Healey, K. 1993. *Map series of South America.* International Travel Map Productions, Van-couver. Five sheets at a scale of 1:4,000,000.

Houseal, B. 1994. Critical parks and reserves proposed and unprotected sites, existing adequate protected areas. Latin American region. The Nature Conservancy "Parks in Peril" Pro-gram, The Nature Conservancy, Washington, D.C. Unpublished document.

IGACC. 1989. *Nueva atlas nacional de Cuba.* Instituto de Geografia de la Academia de Ciencias de Cuba, Havana.

INDERENA. 1985. *Parques nacionales y naturales, santuarios de fauna y flora, reservas forestales y especiales.* Draft dyeline map at a scale of 1:1,500,000. INDERENA, Bogotá, Colombia.

INPARQUES. 1994. *Parques nacionales en peligro.* Dirección General Sectorial de Parques Nacionales, Instituto Nacional de Parques, Caracas, Venezuela.

Instituto Ecuatoriano Forestal y de Áreas Naturales y Vida Silvestre. 1990. *Mapa del sistema de áreas naturales protegidas en el Ecuador.* Map at a scale of 1:1,000,000. Instituto Ecuatoriano Forestal y de Áreas Naturales y Vida Silvestre, Quito, Ecuador.

IUCN. 1992. *Protected areas of the World: A review of national systems. Volume 4: Nearctic and Neotropical.* The World Conservation Union, Gland, Switzerland.

Ministerio del Ambiente y de los Recursos Naturales Renovables and SEFORVEN. 1991. *República de Venezuela: Areas bajo regimen de administración especial.* Map (no scale given). Ministerio del Ambiente y de los Recursos Naturales Renovables and SEFORVEN, Caracas, Venezuela.

Morales Kreuzer, I. 1993. *Monitoreo del bosque en el Departamento de Santa Cruz.* Proyecto de protección de los recursos naturales en el Departamento de Santa Cruz. Cordecruz-KFW-Consorcio IP/CES/KWC.

NARMAP and The Belize Zoo. 1994. *Belize: Protected areas.* Inset map on poster (no scale or projection given). NARMAP and The Belize Zoo, Belize.

National Geographic Society. 1992. *Amazonia: a world resource at risk.* Map at a scale of 1:10,650,000; Chamberlin Trimetric Projection. National Geographic Society, Washington, D.C.

OAS. 1992. *Uruguay: estudio ambiental nacional.* Organization of American States, Washington, D.C.

Peres, C.A., and J.W. Terborgh. 1995. Amazonian nature reserves: an analysis of the defensibility status of existing conservation units and design criteria for the future. *Conservation Biology* 9: 34-46.

Perez, M.R., J.A. Sayer, and S. Cohen Jehoram. 1992. El extractivismo en America Latina: conclusiones y recomendaciones del taller UICN-CEE, Amacayacú, Octubre 1992. World Conservation Union (IUCN), Gland, Switzerland. Draft document.

SINAPE. 1994. *Sistema Nacional de Áreas Naturales Protegidas por el Estado.* Protected area map of Peru (no scale or projection given). Dirección General de Desarrollo Turistico, Lima, Peru.

Torres, H. 1992. Biological diversity in South America: Conservation, management and sustainable use. Unpublished document.

Vila, A.R., and C. Bertonatti. 1993. *Situación ambiental de la Argentina: Recomendaciones y prioridades de acción.* Boletin Tecnico 14. Fundación Vida Silvestre Argentina, Buenos Aires.

World Conservation Monitoring Centre. 1994. *Protected areas of Latin America and the Caribbean.* World Conservation Monitoring Centre, Cambridge, U.K. Draft protected areas digital database for LAC region that classifies each protected area according to IUCN and other relevant categories. The database is based on a wide variety of information sources described in the following document: World Conservation Monitoring Centre. 1993. *The WCMC biodiversity map library: Availability and distribution of GIS datasets.* This list has been updated in the course of this study.

WWF and IUCN. 1994. *Centres of plant diversity: a guide for their strategy and conservation.* World Conservation Union Publication Unit, Cambridge, U.K.

Glossary

Alpha diversity	Species diversity within a habitat.
ARC/INFO	GIS software package developed by Environmental Systems Research Institute (ESRI).
Beta diversity	Species diversity between habitats (thus reflecting changes in species assemblages along environmental gradients).
Biodiversity conservation priority	Four classes of biodiversity conservation priority were determined in this study by overlaying biological distinctiveness on conservation status (see matrix in Chapter 1). The different classes, from Highest Priority at Regional Scale (I) to Important at National Scale (IV), roughly reflect the concern with which we should view the erosion of biodiversity in different ecoregions (Map 8) and the timing and sequence of response by governments and donors to halt the loss of biodiversity.
Biological distinctiveness	Scale-dependent assessment of the biological importance of an ecoregion based on species richness, endemism, relative scarcity of ecoregion type, and rarity of ecological phenomena. Biological distinctiveness classes are Globally Outstanding, Regionally Outstanding, Bioregionally Outstanding, and Locally Important (Map 7).
Bioregion	One of nine biogeographic divisions of LAC, consisting of contiguous ecoregions, designed to better address the biogeographic distinctiveness of ecoregions (Map 1).
Bioregional representation	When designating the biodiversity conservation priority for each ecoregion, bioregional representation is achieved by ensuring that, in almost all cases, for each major habitat type in each bioregion, at least one ecoregion classed Highest Priority at Regional Scale (level I) is designated. When no ecoregion has been classed level I, a single level II (or under some circumstances, a level III) is therefore elevated to a "Iª status" (Map 9).
Bioregionally Outstanding	Biological distinctiveness class.

Conservation status	Assessment of the status of ecological processes and of the viability of species populations in an ecoregion. The different status categories used are Extinct, Critical, Endangered, Vulnerable, Relatively Stable, and Relatively Intact. The snapshot conservation status (Map 5) is based on an index derived from values of five landscape-level variables. The final conservation status (Map 6) is the snapshot assessment modified by an analysis of threats to the ecoregion over the next 20 years.
Conversion	Habitat that is no longer "intact" is considered to be converted. The rate of conversion is a landscape-level variable that measures the percentage of intact habitat in the ecoregion being converted per annum.
Critical	Conservation status category one level below "Extinct" and characterized by low probability of persistence of remaining intact habitat.
Degradation	Landscape-level variable used only for two major habitat types (grasslands, savannas, and shrublands, and montane grasslands) in which it was deemed particularly important and for which data were available.
Digital Chart of the World	Comprehensive 1:1,000,000 base map of the world with multiple data layers and accessible using GIS software.
Ecological processes	Complex mix of interactions between animals, plants and their environment that ensure that an ecosystem's full range of biodiversity is adequately maintained. Examples include population and predator-prey dynamics, pollination and seed dispersal, nutrient cycling, migration, and dispersal.
Ecoregion	Geographically distinct assemblage of natural communities that (a) share a large majority of their species and ecological dynamics; (b) share similar environmental conditions; and (c) interact ecologically in ways that are critical for their long-term persistence. In this study, 191 ecoregions (including 13 mangrove complexes) are defined (Map 3 and large-format map).
Endangered	Conservation status category between Critical and Vulnerable and characterized by medium to low probability of persistence of remaining intact habitat.
Endemism	Degree to which a geographically circumscribed area, such as an ecoregion or a country, contains species not naturally occurring elsewhere.
Extinct	Habitually used for a species or population that has been lost. In this study, also a conservation status category used for an ecoregion with no natural communities resembling original ecosystems remaining.
Extirpated	Status of a species or population that has completely vanished from a given area but which continues to exist in some other location.
Fragmentation	Landscape-level variable measuring the degree to which remaining habitat is separated into smaller discrete blocks.
Gallery forest	Narrow strips of forest along the margins of rivers in otherwise unwooded landscapes.

Gamma diversity — Species diversity at a regional scale (thus reflecting changes in species assemblages over relatively large geographic areas).

Globally Outstanding — Biological distinctiveness class.

Guild — Group of organisms, not necessarily taxonomically related, that are ecologically similar in characteristics such as diet, behavior, or microhabitat preference, or with respect to their ecological role in general.

Habitat blocks — Landscape-level variable that assesses the number and extent of blocks of contiguous habitat, taking into account size requirements for populations and ecosystems to function naturally. It is measured here by a habitat-dependent and ecoregion size-dependent system.

Habitat loss — Landscape-level variable that refers to the percentage of the original land area of the ecoregion that has been lost (converted). It underscores the rapid loss of species and disruption of ecological processes predicted to occur in ecosystems when the total area of remaining habitat declines.

Intact habitat — Relatively undisturbed areas characterized by the maintenance of most original ecological processes and by communities with most of their original native species still present.

Keystone species — Species which are critically important for maintaining ecological processes or diversity of their ecosystems (e.g., large predators, such as the jaguar, in some tropical forest ecosystems).

Landscape ecology — Branch of ecology concerned with the relationship between landscape-level features, patterns, and processes and the conservation and maintenance of ecological processes and biodiversity in entire ecosystems.

Landscape-level variables — Parameters used in this study to assess the conservation status of an ecoregion. The variables used are habitat loss, habitat blocks, fragmentation, conversion, degradation (for some ecoregion types), and protection.

Locally Important — Biological distinctiveness class.

Major ecosystem type — Set of major habitat types whose ecoregions (a) share comparable ecosystem dynamics; (b) have similar response characteristics to disturbance; (c) exhibit similar degrees of beta diversity; and (d) require an ecosystem-specific conservation approach. Five major ecosystem types (METs) are defined in this study.

Major habitat type — Set of ecoregions that (a) experience comparable climatic regimes; (b) have similar vegetation structure; (c) display similar spatial patterns of biodiversity; and (d) contain flora and fauna with similar guild structures and life histories. Eleven major habitat types (MHTs) are defined in this study.

Mangrove complex — Mangroves are a salt-tolerant ecosystem that occupies sheltered tropical and subtropical coastal estuarine environments. In this study we define 13 major biogeographic units called mangrove complexes and subdivide these in turn into a total of 40 smaller mangrove units (Map 4).

Palmar	Palm-dominated savanna of Mexico.
Paramo	Distinctive grassland/savanna habitats found above treeline in the Andes and in Costa Rica.
Protection	Landscape-level variable that assesses how well humans have conserved large blocks of intact habitat and the biodiversity they contain. It is measured here by the number of protected blocks and their sizes in a habitat-dependent and ecoregion size-dependent system.
Red Data system	IUCN has developed and pioneered the use of categories to classify species according to the degree to which they are threatened. The present report adapts this system to ecoregions.
Regionally Outstanding	Biological distinctiveness class.
Relatively Intact	Conservation status category indicating the least possible disruption of ecosystem processes. Natural communities are largely intact with species and ecosystem processes occurring within their natural ranges of variation.
Relatively Stable	Conservation status category between Vulnerable and Relatively Intact in which extensive areas of intact habitat remain but in which local species declines and disruptions of ecological processes have occurred.
Restinga	Distinctive coastal dune habitats found in some areas of northern and eastern South America and in some areas of Mexico.
Vulnerable	Conservation status category between Endangered and Relatively Stable and characterized by good probability of persistence of remaining intact habitat (assuming adequate protection) but also by loss of some sensitive or exploited species.
Xeric	Xeric habitats are dryland and desert areas.
Zacatonal	High altitude alpine tundra formations that occur on several large volcanoes in the Transvolcanic range of central Mexico.

References

Berger, J. 1990. Persistence of different-sized populations: An empirical assessment of rapid extinction in Bighorn Sheep. *Conservation Biology* 4: 91-98.

BSP/CI/TNC/WRI/WWF. 1995. *A regional analysis of geographic priorities for biodiversity conservation in LAC*. A report for USAID. Biodiversity Support Program, Washington, D.C.

Cajal, J.L. 1994. *Programa integrado de conservación ambiental y desarrollo sustentable de la Cordillera de los Andes*. Fundación para la Conservación de las Especies y el Medio Ambiente (FUCEMA), Buenos Aires.

Cintrón, G., A.E. Lugo, D.J. Pool, and G. Morris. 1978. Mangroves of arid environments in Puerto Rico and adjacent islands. *Biotropica* 10: 110-121.

Cintrón, G., and Y. Schaeffer-Novelli. 1985. Caracteristicas y desarrollo estructural de los manglares de Norte y Sur America. *Ciencia Interamericana* (OAS) 25: 4-15.

Collar, N.J., L.P. Gonzaga, N. Krabbe, A. Madroño Nieto, L.G. Naranjo, T.A. Parker, III, and D.C. Wege. 1992. *Threatened birds of the Americas: The ICBP/IUCN Red Data Book*. 3rd edition. International Council for Bird Preservation, Cambridge, U.K.

Craighead, F.C., and V.C. Gilbert. 1962. The effects of hurricane Donna on the vegetation of southern Florida. *Quarterly Journal of the Florida Academy of Science* 25: 1-28.

Csuti, B. In prep. *Gap analysis: identification of priority areas for biodiversity management and conservation*. Draft report in preparation.

Dinerstein, E., V. Krever, D.M. Olson, and L. Williams. 1994. An emergency strategy to rescue Russia's biological diversity. *Conservation Biology* 8: 934-942.

Galloway, W.E. 1975. Process framework for describing the morphologic and stratigraphic evolution of depositional environments. Pages 87-98 in M.L. Broussard, editor, *Deltas: Models for exploration*. Houston Geological Society, Houston.

Groom, M.J., and N. Shumaker. 1993. Evaluating landscape change: Patterns of worldwide deforestation and local fragmentation. Pages 22-44 in P.M. Kareiva, J.M. Kingsolver, and R.B. Huey, editors, *Biotic interactions and global change*. Sinauer Associates, Inc., Sunderland, Massachusetts.

Hayes, M.O. 1975. Morphology of sand accumulations in estuaries. Pages 3-22 in L.E. Cronin, editor, *Estuarine Research, vol. 2. Geology and Engineering*. Academic Press, New York.

Heald, E.J. 1969. *The production of organic detritus in a south Florida estuary*. Ph.D. diss. University of Miami.

IUCN (World Conservation Union). 1988. *IUCN red list of threatened animals*. IUCN, Gland, Switzerland.

Kareiva, P., and U. Wennergren. 1995. Connecting landscape patterns to ecosystem and population processes. *Nature* 237: 299-302.

Kelleher, G., C. Bleakley, and S. Wells, principal editors. 1995. *A global representative system of marine protected areas*. Four volumes. The Great Barrier Reef Marine Park Authority, The World Bank, and The World Conservation Union (IUCN), Washington, D.C.

Krever, V., E. Dinerstein, D.M. Olson, and L. Williams, editors. 1994. *Conserving Russia's biological diversity: An analytical framework and initial investment portfolio*. Conservation Science Publication Series. No. 1. World Wildlife Fund, Washington, D.C.

Laurance, W.F. 1991. Ecological correlates of extinction proneness in Australian tropical rain forest mammals. *Conservation Biology* 5: 79-89.

Long, A.J., M.J. Crosby, A.J. Stattersfield, and D. Wege. 1994. *Towards a global map of biodiversity:*

Patterns in the distribution of restricted-range birds. Birdlife International, Cambridge, U.K. Unpublished document.

Lovejoy, T.E. 1980. Discontinuous wilderness: Minimum areas for conservation. *Parks* 5: 13-15.

Lugo, A.E., and C. Patterson-Zucca. 1977. The impact of frost stress on mangrove ecosystems. *Tropical Ecology* 18: 149-61.

Lugo, A.E., and S.C. Snedaker. 1974. The ecology of mangroves. *Annual Review of Ecology and Systematics* 5: 39-64.

Mace, G., and R. Lande. 1991. Assessing extinction threats: Toward a reevaluation of IUCN threatened species categories. *Conservation Biology* 5: 148-157.

Margules, C.R., A.O. Nicholls, and R.L. Pressey. 1988. Selecting networks of reserves to maximize biological diversity. *Biological Conservation* 43: 63-76.

Meffe, G.K., and C.R. Carroll. 1994. *Principles of conservation biology.* Sinauer Associates, Inc., Sunderland, Massachusetts.

Mittermeier, R., and T. Werner. 1990. Wealth of plants and animals unites "megadiversity" countries. *Tropicus* 4: 4-5.

Myers, N. 1988. Threatened biotas: Hotspots in tropical forests. *The Environmentalist* 8: 178-208.

Newmark, W.D. 1991. Tropical forest fragmentation and the local extinction of understory birds in the eastern Usambara Mountains, Tanzania. *Conservation Biology* 5: 67-78.

Noss, R.F. 1992. The wildlands project and conservation strategy. *Wild Earth*, special issue: 10-25.

Noss, R.F., and A.Y. Cooperrider 1994. *Saving nature's legacy: Protecting and restoring biodiversity.* Island Press, Washington, D.C.

Odum, W.E. 1971. *Pathways of energy flow in a South Florida estuary.* University of Miami Sea Grant Bulletin No. 7.

Olson, D.M., D. DellaSala, E. Dinerstein, S.A. Primm, and M. Forney. 1994. *Identifying a system of core reserves for the inland forests of the Pacific Northwest: A landscape ecology approach.* Draft document submitted to the Eastside Ecosystem Management Team, USDA Forest Service.

Olson, D.M., and E. Dinerstein. 1994. *Assessing the conservation potential and degree of threat among ecoregions of Latin America and the Caribbean: A proposed landscape ecology approach.* LATEN Dissemination Note No. 10. The World Bank, Latin America Technical Department, Environment Division.

Pool, D.J., S.C. Snedaker, and A.E. Lugo. 1977. Structure of mangrove forests in Florida, Puerto Rico, Mexico and Costa Rica. *Biotropica* 9: 195-212.

Pressey, R.L., C. J. Humphries, C.R. Margules, R.I. Vane-Wright, and P.H. Williams. 1993. Beyond opportunism: Key principles for systematic reserve selection. *Trends in Ecology and Evolution* 8: 124-128.

Primack, R.B. 1993. *Essentials of conservation biology.* Sinauer Associates, Inc., Sunderland, Massachusetts.

Redford, K.H., A. Taber, and J.A. Simonetti. 1990. There is more to biodiversity than tropical rain forests. *Conservation Biology* 4: 328-330.

Saunders, D.A., R.J. Hobbs, and C.R. Margules. 1991. Biological consequences of ecosystem fragmentation: A review. *Conservation Biology* 5: 18-32.

Schaeffer-Novelli, Y., G. Cintrón, R. Adaime, and T. de Camargo. 1990. Variability of the mangrove ecosystem along the Brazilian coast. *Estuaries* 13: 204-219.

Scott, M. et al. 1993. Gap analysis: A geographic approach to protection of biological diversity. *Wildlife Monographs* 123: 1-4.

Shafer, C.L. 1995. Value and shortcomings of small reserves. *BioScience* 45: 80-88.

Simberloff, D. 1992. Do species-area curves predict extinction in fragmented forest? Pages 75-89 in T.C. Whitmore and J.A. Sayer, editors, *Tropical deforestation and species extinction.* Chapman and Hall, London.

Skole, D.L., and C. Tucker. 1993. Tropical deforestation and habitat fragmentation in the Amazon: Satellite data from 1978 to 1988. *Science* 260: 1905-1910.

Tabb, D.C., and A.C. Jones. 1962. Effects of hurricane Donna on the aquatic fauna of North Florida Bay. *Transactions of the American Fisheries Society* 91: 375-378.

Terborgh, J. 1992. The maintenance of diversity in tropical forests. *Biotropica* 24: 283-292.

Thom, B.G. 1967. Mangrove ecology and deltaic geomorphology: Tabasco, Mexico. *Journal of Ecology* 55: 301-43.

___ 1984. Coastal landforms and geomorphic processes. Pages 3-17 in S.C. Snedaker and J.G. Snedaker, editors, *The mangrove ecosystem: Research methods.* UNESCO, Paris.

Twilley, R.R. 1982. *Litter dynamics and organic carbon exchange in black mangrove (Avicennia germinans) basin forests in a southwest Florida estuary.* Ph.D. diss. University of Florida, Gainesville.

Wege, D.C., and A. Long, editors. In press. *Priority areas for threatened birds in the Neotropics.* Bird Conservation Series No. 5. Birdlife International, Cambridge, U.K.

Whitmore, T.C., and J.A. Sayer, editors. 1992. *Tropical deforestation and species extinction.* Chapman and Hall, London.

Wilcove, D.S., C.H. McLellan, and A.P. Dobson. 1986. Habitat fragmentation in the temperate zone. Pages 237-256 in M.E. Soule, editor, *Conservation biology: The science of scarcity and diversity.* Sinauer Associates, Inc., Sunderland, Massachusetts.

Wilcox, B.A., and K.N. Duin. 1995. Indigenous cultural diversity and biological diversity: Overlapping values of Latin American ecoregions. *Cultural Survival Quarterly*, Winter.

Wood, D., and S. Cornelius. 1994. *Biogeographic priorities for the Latin America and Caribbean Program.*

World Wildlife Fund, Washington, D.C. Unpublished document.

Woodwell, G.M. 1995. A habitat for all life. *BioScience* 45: 67-68.

WWF (World Wildlife Fund) and IUCN (World Conservation Union). 1994. *Centers of plant diversity: A guide and strategy for their conservation.* IUCN Publications Unit, Cambridge, U.K.

Map 1. Bioregions of Latin America and the Caribbean

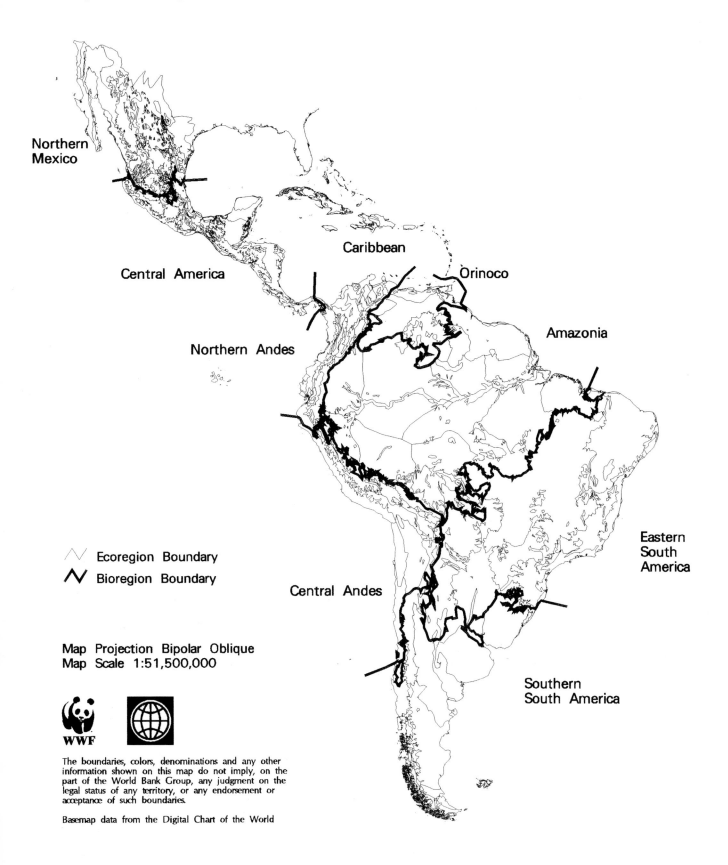

Northern
Mexico

Central America

Caribbean

Orinoco

Amazonia

Northern Andes

Ecoregion Boundary

Bioregion Boundary

Eastern
South
America

Central Andes

Map Projection Bipolar Oblique
Map Scale 1:51,500,000

Southern
South
America

WWF

The boundaries, colors, denominations and any other
information shown on this map do not imply, on the
part of the World Bank Group, any judgment on the
legal status of any territory, or any endorsement or
acceptance of such boundaries.

Basemap data from the Digital Chart of the World

Map 2a. Major Habitat Types of Mexico and Central America

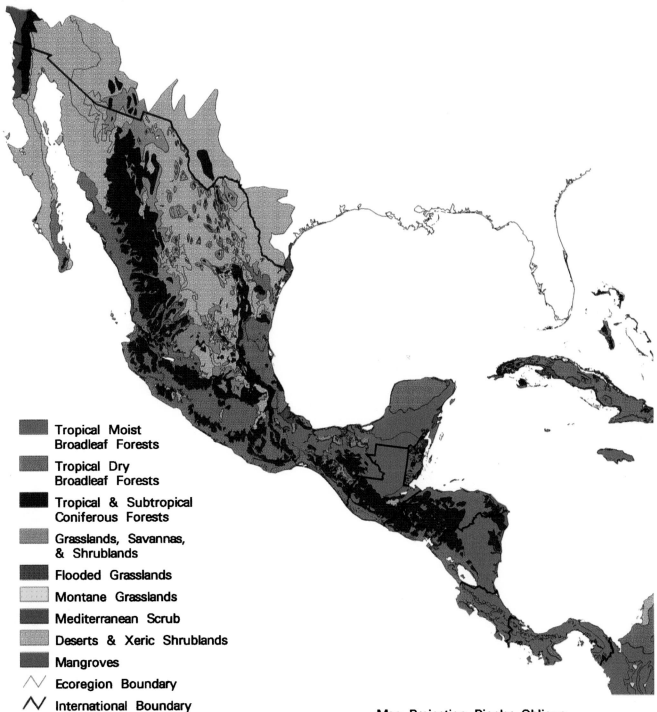

Tropical Moist
Broadleaf Forests

Tropical Dry
Broadleaf Forests

Tropical & Subtropical
Coniferous Forests

Grasslands, Savannas,
& Shrublands

Flooded Grasslands

Montane Grasslands

Mediterranean Scrub

Deserts & Xeric Shrublands

Mangroves

/\/ Ecoregion Boundary

/\/ International Boundary

Map Projection Bipolar Oblique
Map Scale 1:22,000,000

WWF

Map 2b. Major Habitat Types of the Caribbean

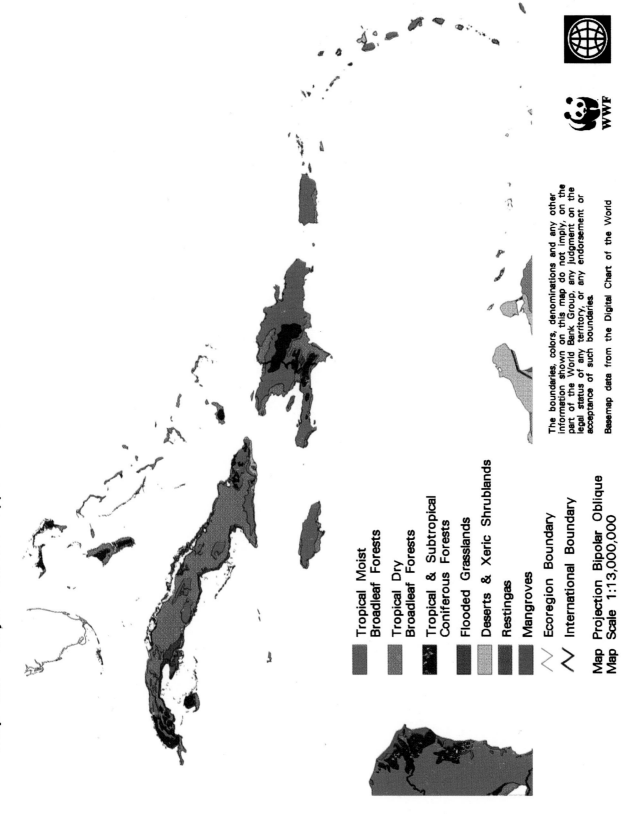

Tropical Moist
Broadleaf Forests

Tropical Dry
Broadleaf Forests

Tropical & Subtropical
Coniferous Forests

Flooded Grasslands

Deserts & Xeric Shrublands

Restingas

Mangroves

Ecoregion Boundary

International Boundary

Map Projection Bipolar Oblique
Map Scale 1:13,000,000

The boundaries, colors, denominations and any other
information shown on this map do not imply, on the
part of the World Bank Group, any judgment on the
legal status of any territory, or any endorsement or
acceptance of such boundaries.

Basemap data from the Digital Chart of the World

Map 2c. Major Habitat Types of South America

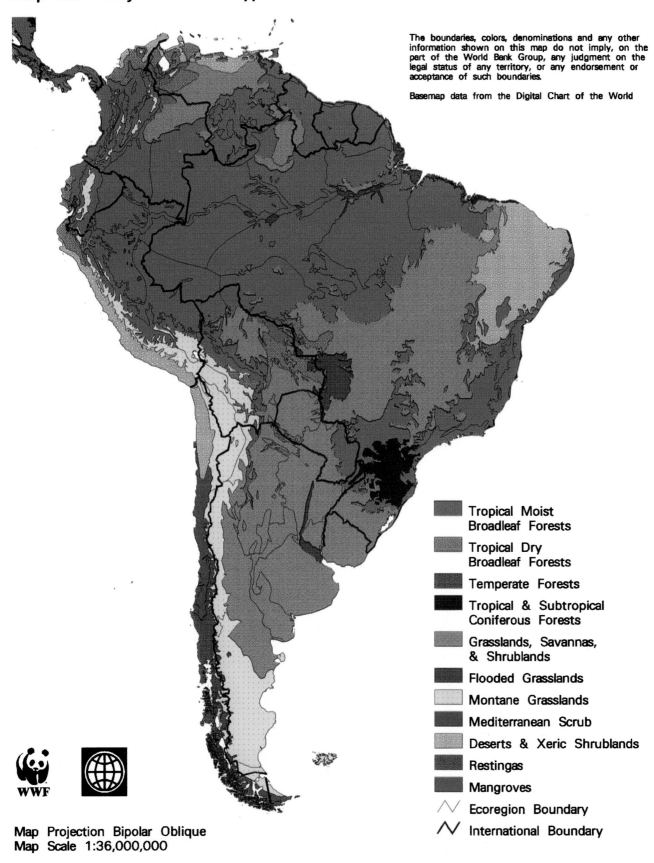

Tropical Moist
Broadleaf Forests

Tropical Dry
Broadleaf Forests

Temperate Forests

Tropical & Subtropical
Coniferous Forests

Grasslands, Savannas,
& Shrublands

Flooded Grasslands

Montane Grasslands

Mediterranean Scrub

Deserts & Xeric Shrublands

Restingas

Mangroves

Ecoregion Boundary

International Boundary

WWF

Map Projection Bipolar Oblique
Map Scale 1:36,000,000

Map 3. Ecoregions of Latin America and the Caribbean

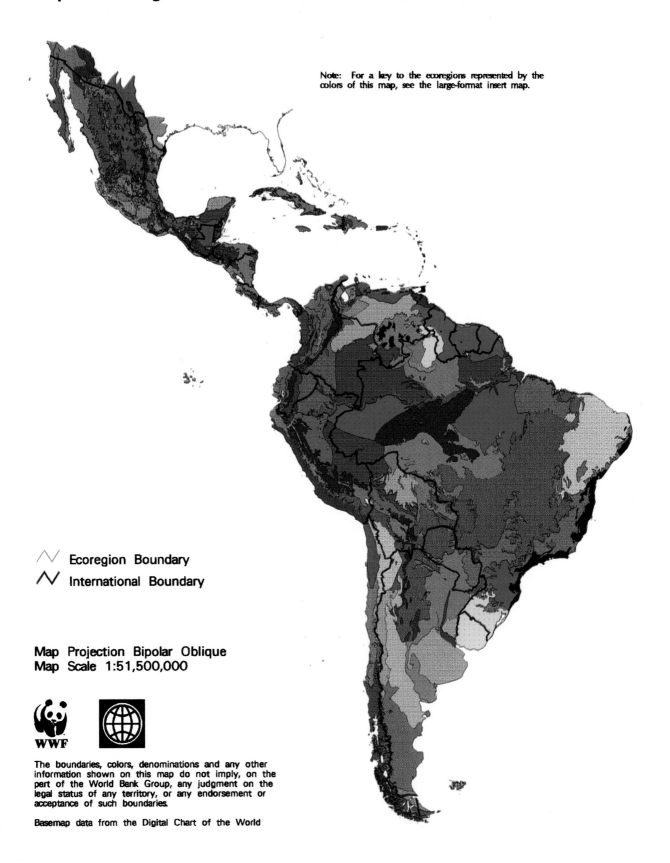

Note: For a key to the ecoregions represented by the colors of this map, see the large-format insert map.

/\/ Ecoregion Boundary
/\/ International Boundary

Map Projection Bipolar Oblique
Map Scale 1:51,500,000

WWF

The boundaries, colors, denominations and any other information shown on this map do not imply, on the part of the World Bank Group, any judgment on the legal status of any territory, or any endorsement or acceptance of such boundaries.

Basemap data from the Digital Chart of the World

Map 4. Mangrove Complexes and Units of Latin America and the Caribbean

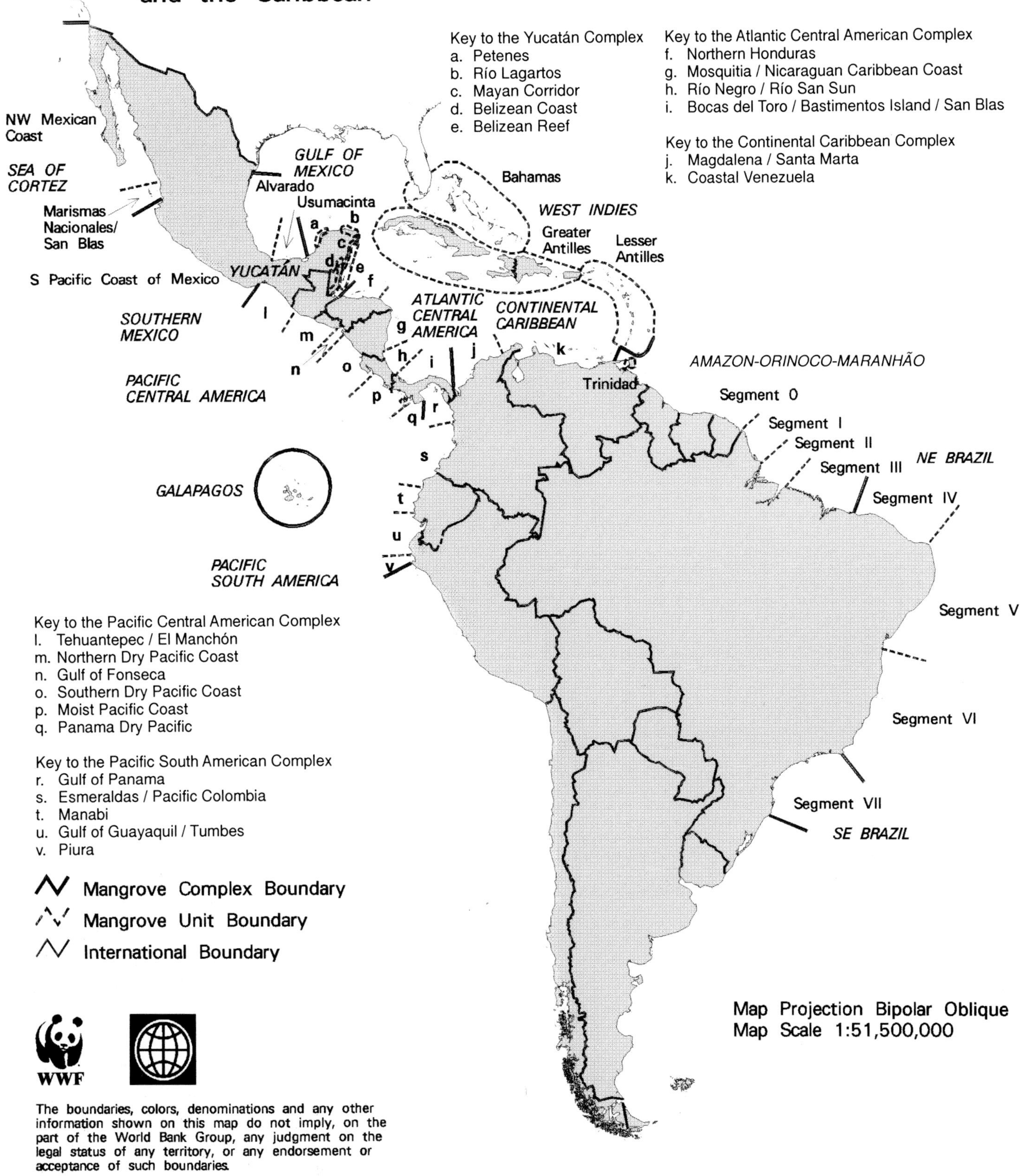
Key to the Yucatán Complex
a. Petenes
b. Río Lagartos
c. Mayan Corridor
d. Belizean Coast
e. Belizean Reef
Key to the Atlantic Central American Complex
f. Northern Honduras
g. Mosquitia / Nicaraguan Caribbean Coast
h. Río Negro / Río San Sun
i. Bocas del Toro / Bastimentos Island / San Blas
Key to the Continental Caribbean Complex
j. Magdalena / Santa Marta
k. Coastal Venezuela
NW Mexican Coast
SEA OF CORTEZ
Marismas Nacionales/ San Blas
S Pacific Coast of Mexico
GULF OF MEXICO
Alvarado
Usumacinta
YUCATÁN
SOUTHERN MEXICO
PACIFIC CENTRAL AMERICA
ATLANTIC CENTRAL AMERICA
CONTINENTAL CARIBBEAN
Bahamas
WEST INDIES
Greater Antilles
Lesser Antilles
Trinidad
AMAZON-ORINOCO-MARANHÃO
Segment 0
Segment I
Segment II
Segment III
NE BRAZIL
Segment IV
Segment V
Segment VI
Segment VII
SE BRAZIL
GALAPAGOS
PACIFIC SOUTH AMERICA
Key to the Pacific Central American Complex
l. Tehuantepec / El Manchón
m. Northern Dry Pacific Coast
n. Gulf of Fonseca
o. Southern Dry Pacific Coast
p. Moist Pacific Coast
q. Panama Dry Pacific
Key to the Pacific South American Complex
r. Gulf of Panama
s. Esmeraldas / Pacific Colombia
t. Manabi
u. Gulf of Guayaquil / Tumbes
v. Piura
Mangrove Complex Boundary
Mangrove Unit Boundary
International Boundary
Map Projection Bipolar Oblique
Map Scale 1:51,500,000
WWF

The boundaries, colors, denominations and any other information shown on this map do not imply, on the part of the World Bank Group, any judgment on the legal status of any territory, or any endorsement or acceptance of such boundaries.

Basemap data from the Digital Chart of the World

Map 5. Snapshot Conservation Status of Ecoregions of Latin America and the Caribbean

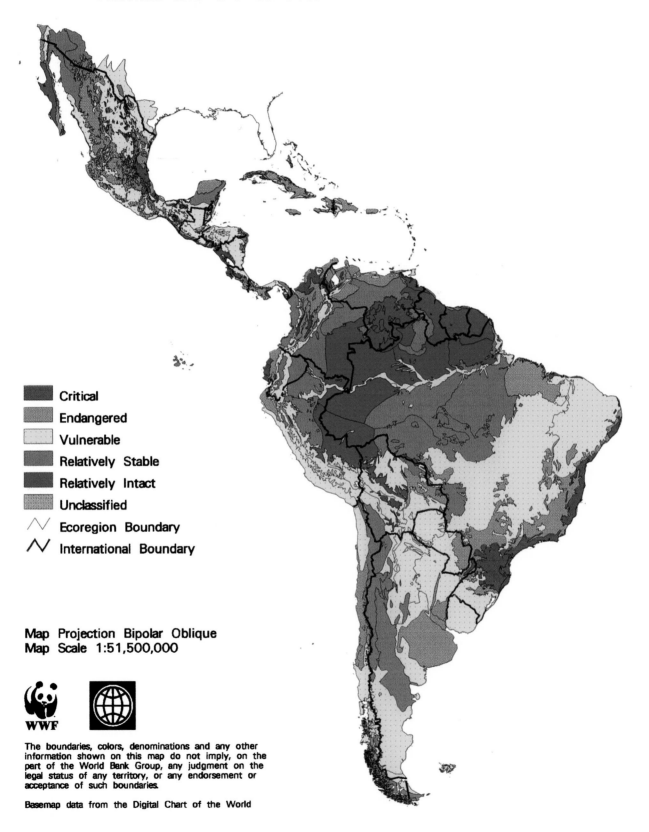

Critical

Endangered

Vulnerable

Relatively Stable

Relatively Intact

Unclassified

⋀ Ecoregion Boundary

⋀ International Boundary

Map Projection Bipolar Oblique
Map Scale 1:51,500,000

WWF

The boundaries, colors, denominations and any other information shown on this map do not imply, on the part of the World Bank Group, any judgment on the legal status of any territory, or any endorsement or acceptance of such boundaries.

Basemap data from the Digital Chart of the World

Map 6. Final Conservation Status of Ecoregions of Latin America and the Caribbean (Snapshot Conservation Status Modified by Threat)

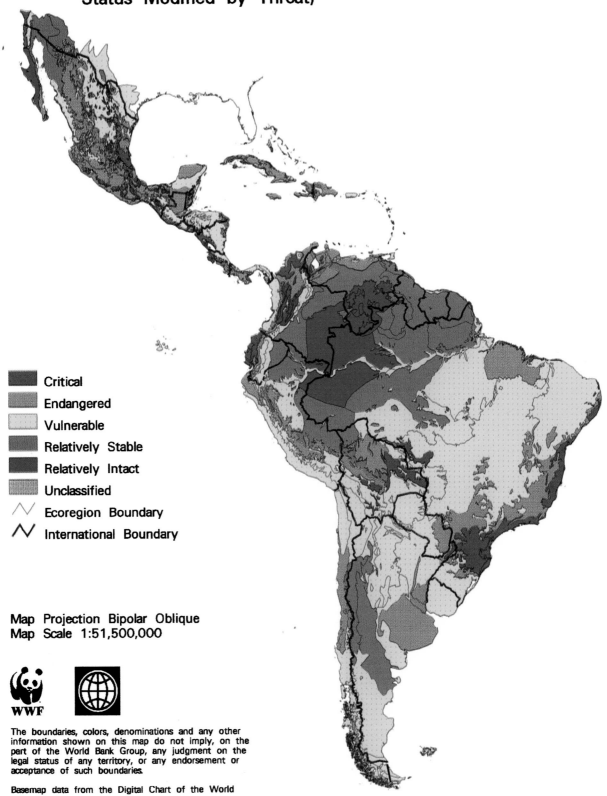

Critical

Endangered

Vulnerable

Relatively Stable

Relatively Intact

Unclassified

Ecoregion Boundary

International Boundary

Map Projection Bipolar Oblique
Map Scale 1:51,500,000

WWF

Map 7. Biological Distinctiveness of Ecoregions of Latin America and the Caribbean

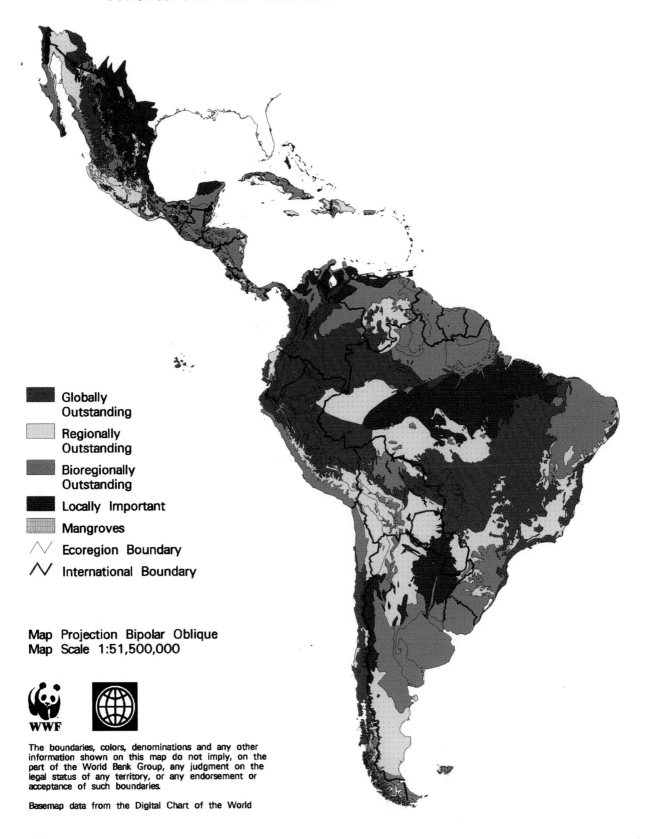

Globally Outstanding

Regionally Outstanding

Bioregionally Outstanding

Locally Important

Mangroves

∿ Ecoregion Boundary

∧ International Boundary

Map Projection Bipolar Oblique
Map Scale 1:51,500,000

WWF

Basemap data from the Digital Chart of the World

Map 8. Biodiversity Conservation Priority of Ecoregions of Latin America and the Caribbean

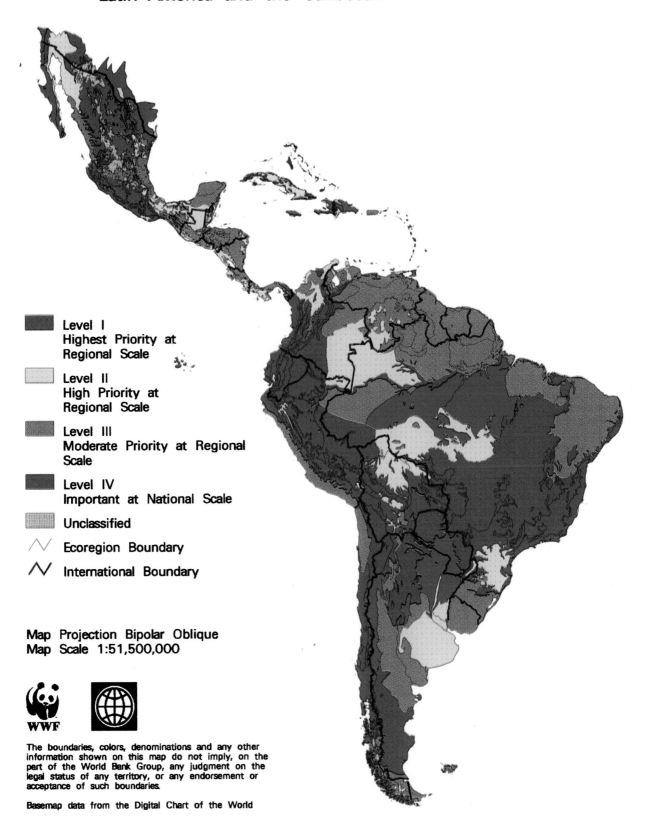

Level I
Highest Priority at
Regional Scale

Level II
High Priority at
Regional Scale

Level III
Moderate Priority at Regional
Scale

Level IV
Important at National Scale

Unclassified

Ecoregion Boundary

International Boundary

Map Projection Bipolar Oblique
Map Scale 1:51,500,000

WWF

Map 9. Biodiversity Conservation Priority of Ecoregions of Latin America and the Caribbean (Incorporating Consideration of Bioregional Representation)

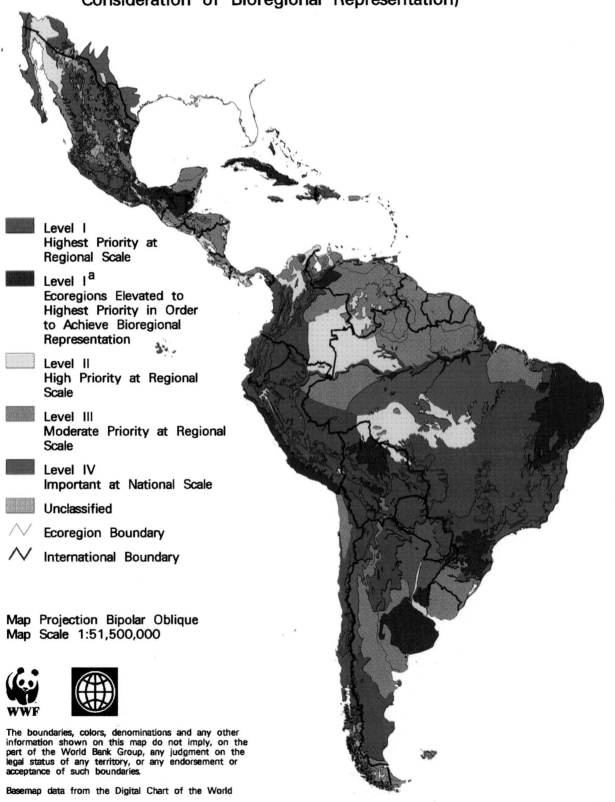

Level I
Highest Priority at
Regional Scale

Level I[a]
Ecoregions Elevated to
Highest Priority in Order
to Achieve Bioregional
Representation

Level II
High Priority at Regional
Scale

Level III
Moderate Priority at Regional
Scale

Level IV
Important at National Scale

Unclassified

Ecoregion Boundary

International Boundary

Map Projection Bipolar Oblique
Map Scale 1:51,500,000

WWF

Basemap data from the Digital Chart of the World